AF541273

Elasticity of Materials
An Introduction

Elasticity of Materials
An Introduction

– Author –

Nilan Bodapati

Irumporai Dubagunta

ISBN: 9789389183771 (Hardback)

Published by : **WORDPEN ACADEMICS**
4736/23, Ansari Road, Darya Ganj,
New Delhi – 110 002
Phone: +91-011-35004336
E-mail: info@wordpenacademics.com

Printed at **:** **Replika Press Pvt. Ltd.**

Exclusive Distributor: ***Bio-Green Books, New Delhi***

Preface

In physics and materials science, elasticity is the ability of a body to resist a distorting influence and to return to its original size and shape when that influence or force is removed. Solid objects will deform when adequate loads are applied to them; if the material is elastic, the object will return to its initial shape and size after removal. This is in contrast to plasticity, in which the object fails to do so and instead remain in its deformed state. The physical reasons for elastic behavior can be quite different for different materials. In metals, the atomic lattice changes size and shape when forces are applied (energy is added to the system). When forces are removed, the lattice goes back to the original lower energy state. For rubbers and other polymers, elasticity is caused by the stretching of polymer chains when forces are applied. Hooke's law states that the force required to deform elastic objects should be directly proportional to the distance of deformation, regardless of how large that distance becomes. This is known as perfect elasticity, in which a given object will return to its original shape no matter how strongly it is deformed. This is an ideal concept only; most materials which possess elasticity in practice remain purely elastic only up to very small deformations, after which plastic (permanent) deformation occurs. In engineering, the elasticity of a material is quantified by the elastic modulus such as the Young's modulus, bulk modulus or shear modulus which measure the amount of stress needed to achieve a unit of strain; a higher modulus indicates that the material is harder to deform. The SI unit of this modulus is the pascal (Pa). The material's elastic limit or yield strength is the maximum stress that can arise before the onset of plastic deformation. Its SI unit is also the pascal (Pa).

It is a comprehensive resource that covers a range of topics related to elasticity, including theory, applications, and experimental techniques. Provides an introductory chapter that covers analytical and numerical approaches used in engineering elasticity, making it a useful resource for students and professionals alike. Offers a comprehensive overview of stress-strain analysis for elasticity equations, including the use of Hooke's Law, Young's Modulus, and Poisson's Ratio. Covers the use of Finite Element Analysis (FEA) and experimental techniques for determining the applied elasticity problem, with a focus on fabricating aspheric surfaces. Discusses the concept of phase transition based on elastic systematics, exploring how changes in temperature and pressure can cause materials to undergo a transition from elastic to a non-elastic state. Describes the repair inspection technique based on elastic-wave tomography, providing valuable insights into non-destructive testing methods for detecting damage in concrete structures. The book is well-suited for students, researchers, and professionals in engineering and materials science, as well as anyone interested in understanding the behavior of materials under stress and strain.

We are grateful to all those persons as well as various books, manuals, periodicals, magazines, journals etc. that helped in the preparation of this book. In spite of the best efforts, it is possible that some errors may have occurred into the compilation and editing of the book. Further queries, constructive suggestions and criticisms for the improvement of the book are always welcomed and shall be thankfully acknowledged.

Nilan Bodapati
Irumporai Dubagunta

Contents

Section 1

General Theorems in Elasticity

Chapter 1

Introductory Chapter: Analytical and Numerical Approaches in Engineering Elasticity

1. Introduction

In this section, the historical development of the "elasticity theory" was presented briefly, and recent studies performed about the elasticity concept were categorized and listed according to their basic engineering problem groups. The mentioned literature survey has been performed by searching the keywords "elasticity," "analytic," and "solution" between the years 2014 and 2018. The most important general aspects of the "elasticity theory" were described in four groups as "the unknowns," "the used equations," "the modeling procedures," and "solution methods." In the future, in the consideration of these explained theoretical, numerical, and experimental properties, the researchers can be concentrating on the origin of the problem and new solution methods in deciding the exact nature of the material.

The elasticity concept of solid materials is the deformation with the external force application and recovery to its original shape after the forces removed. In the strength measurement of the material, stress (force per area) and strain (deformation per unit length) criteria have been used. The elasticity theory was presented in order to explain the basic theoretical concepts and their analytical solution methods, the deformations that were assumed to be very small and corresponding stress distributions. The classical elasticity theory was explained by theorems of "uniqueness of solution" and "existence of solution" as they have been declared in the basic mathematical concepts. The "uniqueness of solution" theorem was restricted to a single solution space by satisfying the related boundary or the initial conditions. If there were no any boundary or initial conditions, the solution space would have to be infinity. The "existence of solution" theorem was created by explaining the default displacement functions, checking the equilibrium equations for stress definitions, and satisfying the partial differential equations with the infrastructure of the default solutions. The purpose of the elasticity theory was the determination of this unique and exact solution in elastic region of the material. In linear elastic region, superposition method and combined loading applications are widely used in engineering.

2. Historical development in elasticity

The historical development of the concept of "elasticity" by considering mathematics, physics, and engineering mechanics was summarized in **Figure 1** [1, 2]. The scientific studies performed on engineering problems have been grouped as analytical, numerical, and experimental. The main solution techniques listed below

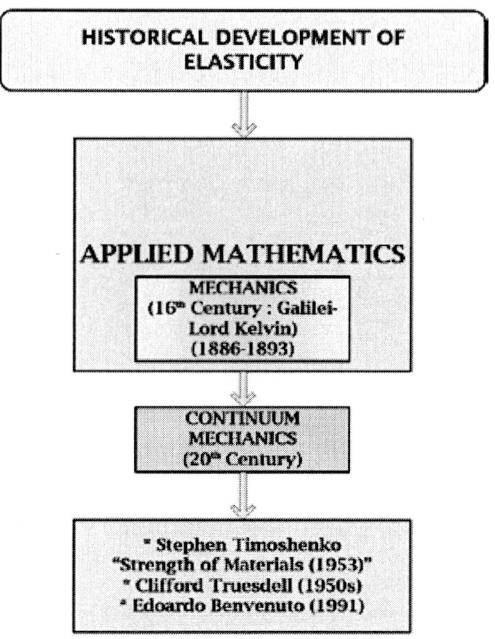

Figure 1.
Development of elasticity between the sixteenth and twentieth centuries.

form the first step aspects in performing the experiments and obtaining the numerical solutions by considering innovations: (i) characteristics of the solution methods, (ii) learning the mathematical theories, (iii) the physics of the problem, and (iv) learning the problem-solving methodologies. The second step aspects have been listed as "solving problems by mathematical techniques" and "obtaining new formulas." The scientific progress has been continued thanks to the studies done since the sixteenth century. The development in scientific area occurred in the elasticity concept has been summarized and visualized in the consideration of the scientists who have lived between the sixteenth and twentieth centuries and their studies [1, 2]. These famous scientists were Galilei (1564–1642), Mariotte (1620–1684), Hooke (1635–1703), Leibniz (1646–1716), Bernoulli (1700–1782), Baumgarten (1706–1757), Euler (1707–1783), Coulomb (1736–1806), Young (1773–1829), Poisson (1781–1840), Navier (1785–1836), Cauchy (1789–1857), Saint-Venant (1797–1886), Borchardt (1817–1880), Rankine (1820–1872), Kirchhoff (1824–1887), Maxwell (1831–1879), Clebsch (1833–1872), Kohlrausch (1840–1910), Amagat (1841–1915), Voigt (1850–1919), Mallock (1851–1933), Lamme (1864–1924), Röntgen (1872,1919), Synge (1897–1995), and Everett (1930–1982) (**Figure 1**).

3. Classification of engineering problems in the context of elasticity

In this section, the results of the literature review on elasticity were evaluated by referring to the articles (total number of articles, 157) between 2014 and 2018. Important information has gained from the literature survey about the elasticity theory and its related recent engineering solutions, as well as information about the theoretical, numerical, and experimental scientific researches and scientific innovations. The brief classification of the main engineering problems was summarized in **Figure 2**.

The studies evaluated in the literature review were listed below in 10 main headings. The distribution of articles corresponding to research concepts is presented in **Figure 3**. These are (1) historical development, (2) analytical and experimental studies related to the finite element method (FEM), (3) experimental studies, (4) analytical studies and finite element analysis (FEA), (5) analytical studies,

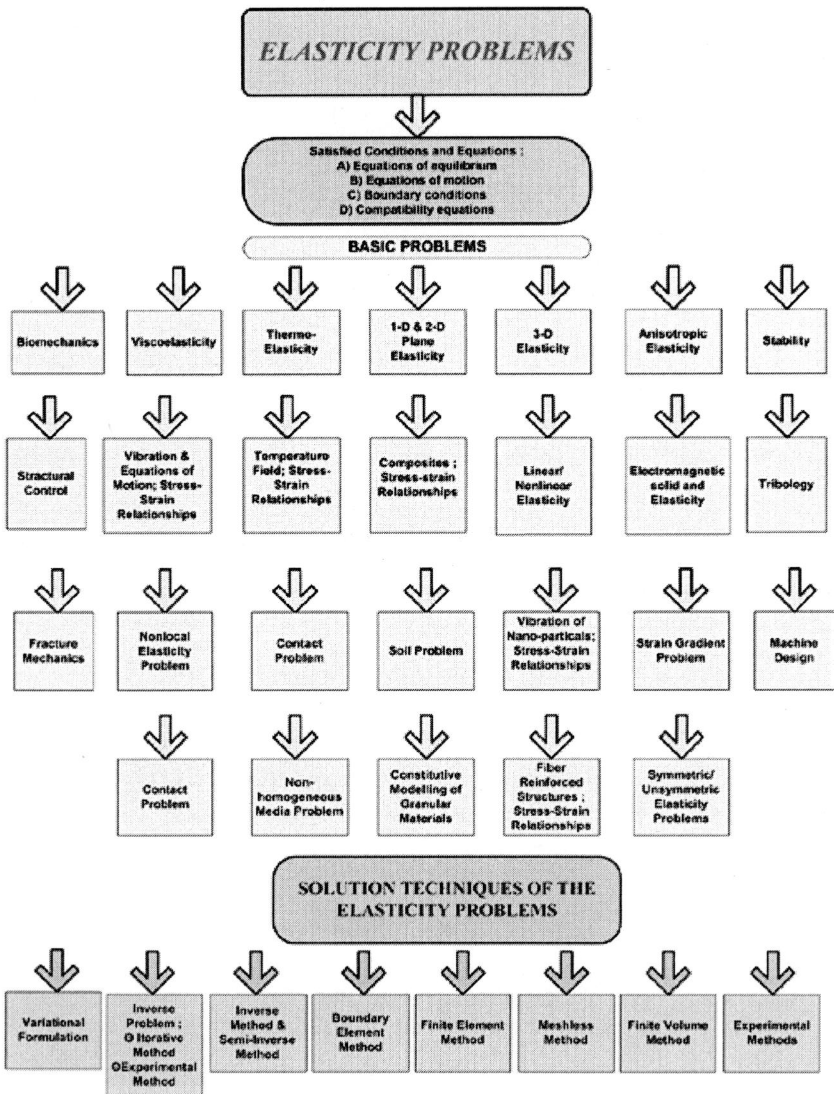

Figure 2.
Classification of the basic elasticity problems and their solution techniques.

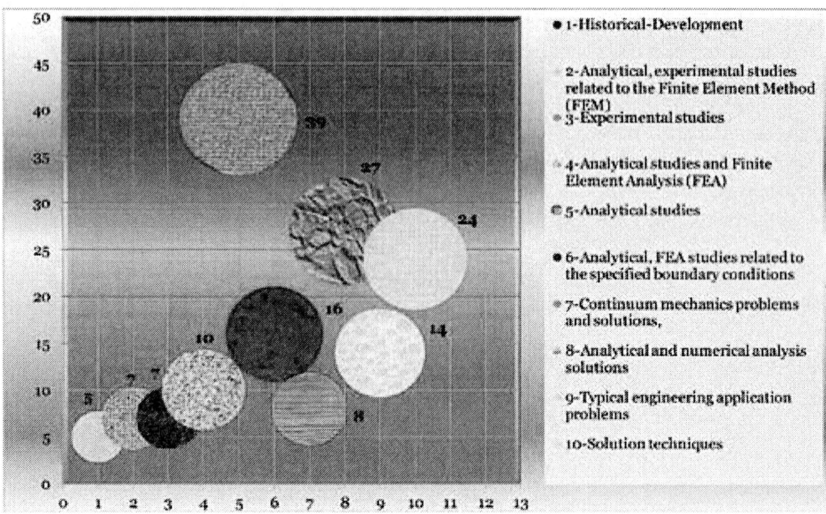

Figure 3.
The results of the literature review on elasticity were evaluated by referring to the 157 articles between 2014 and 2018.

(6) analytical and FEA studies related to the specified boundary conditions, (7) continuum mechanics problems and solutions, (8) analytical and numerical analysis solutions, (9) typical engineering application problems, and (10) solution techniques. The types of elasticity problems have been grouped according to the science innovations and related industrial applications. The numerical problems have been solved in three basic steps. The first step was to check the basic differential equations in terms of satisfaction with the placement of the estimated displacement functions. The second step was to check the "initial values" or the "boundary conditions" of the problem [3–5]. Values were substituted into the differential equations in order to satisfy the conditions at these defined coordinates or at time domains. The boundary conditions have been classified in two groups as "the essential" (displacement) and "the natural" (force) boundary conditions. The initial conditions were the first-stage variations named initiative and time-dependent variables. The third step was the satisfaction of the continuity conditions on the compatibility equations by means of assumed displacement functions. The basic elasticity problems were grouped into 26 subtitles as described in **Figure 2**. In this figure, the number of generally used proposed solution techniques analytically and numerically was equal to eight.

4. General principles in the elasticity theory

Elasticity concept is explainable by the natural elastic behavior of the materials. In elastic region, material deformed in a nonpermanent form up to the elastic limit was reached. The relationship between stress (σ) and strain (ε) under loading and unloading cases was explained by the linear and nonlinear equations. The slopes of the linear curves developed in linear elastic region were known as Young's modulus E, and shear modulus G, of the materials under tensile/compression and torsion tests. During these tests, total calculated area under the linear curves was defined as the total potential energy stored in the material. Proportionally, stress development and strains occurred in the structure according to the applied load. Principally, application of the stress distributions should be very slowly; on the other hand, at each incremental loading step, the equilibrium state and its equilibrium equations of the specimen should be satisfied. This controlled operation and action-reaction principle have worked under the control mechanism of the testing machine. The total work done by incremental external forces "dW" was equal to total potential energy stored incrementally "dU" in the structure of linear elastic region. Using this principle, the governing equations were satisfied by $dW - dU = 0$. Otherwise, in the case which used high strain rates $\dot{\varepsilon}$, the material behavior would have been examined in the material nonlinearity concept. In the nonlinear elastic material experimental tests, the resulting stress-strain curves represented the combination of the behavior of nonlinear continuous or multiple nonlinear continuous forms. In nonlinear curves, the stored potential energy "U" developed in the elastic limit range was calculated in the consideration of two areas: the first area under the $\sigma - \varepsilon$ nonlinear curve described as the stored potential energy by strain increments $\varepsilon + d\varepsilon$ and the second area above the curve, known as the complementary potential energy by stress increments $\sigma + d\sigma$ stored in the material. Both linear and nonlinear elasticity equations were derived according to the assumption that during loading and unloading stages of the experiments, the material stores its potential energy within the molecules and there was no loss of energy. As known in the molecular concept, the binding energy keeps the molecules together at any instant of time, and in the lack of energy loss such as heat or light, there will be no loss in the total mass of the molecular system. This phenomenon shows us that the system, which has no energy loss, does not combine (no binding status) with another solid object or with atoms that oscillates at short distances. Otherwise, in the case of the material decreases in amount as losing its mass as energy in the form of heat or light during the binding process, the removed energy corresponding to the removed mass can be explained by Einstein's equation $E = mc^2$. Here, E is the binding energy, m is the mass change in the system, and c is the speed of light, respectively. The elasticity solutions were grouped in terms of a variety of the material, geometry, and loading types. Generally, the used geometries were selected as bar-, beam-, plate-, and shell-type isotropic or composite-type structures. In order to obtain analytical and numerical solutions, the three-dimensional elasticity problems can be reduced into two-dimensional problems in the consideration of the plane stress and plane strain concepts of the elasticity. By these methods the total number of unknowns will be equal to total numbers of equations. Otherwise, some unknown values will stay in unsolvable or undefined forms. Geometrical, material, and loading symmetries reduce problem-solving difficulties in the analytical and numerical models. On the other hand, continuity conditions in geometries automatically satisfies the continuity conditions in the analytical and numerical solutions of elasticity. For example, the existence of the fourth-order partial derivatives of the assumed solution approximation functions is checking the continuity and compatibility equations. Singularity problems may be discarded by omitting the very small holes, empty spaces, gaps in macroscale, or dislocations and beside these the distances between small particles in microscale. In the case of a three-dimensional problem in elasticity, 15 unknowns were defined as mentioned below. These were six stress components, six strain components, and three displacement components. These unknown values were to be calculated by using 15 elasticity equations, three equilibrium equations, six stress-strain relationships, and six strain-displacement relationships. Continuity conditions were satisfied by considering the six compatibility equations which were derived from 15 elasticity equations in three-dimensional problems. Boundary conditions and the initial conditions were both defined on the boundaries and at the starting time domains, respectively, in order to obtain the solutions under the limitation of approximate and true percentage minimum error calculations. In the case of three-dimensional elasticity problem, 15 unknown values have to be solved by 15 governing equations (the list of the unknowns were six stress components $[\sigma_x \ \sigma_y \ \sigma_z \ \tau_{xy} \ \tau_{yz} \ \tau_{xz}]$ and six strain components $[\varepsilon_x \ \varepsilon_y \ \varepsilon_z \ \gamma_{xy} \ \gamma_{yz} \ \gamma_{xz}]$, and additionally the three displacement components [u v w]) [3, 4]. In solid mechanics and elasticity theory, the governing partial differential equations, the constitutive and kinematics equations, and the initial and boundary conditions have been all defined. However, if at least one of the above conditions has remained partially or entirely unknown, then one has a so-called inverse problem (**Figure 2**) [5]. On the other hand, the elasticity "inverse problem" has been defined for the problems in which they consist of recovering the missing displacements to the solution space corresponding to the applied force data by using the iterative calculation steps. Obviously, lost or uncalculated data developing on one part of a whole domain boundary have directly affected the final configuration of the stress-strain and displacement components and their resultant solution spaces at the other part of this boundary. The proposed solutions were both numerical and analytical (**Figure 2**). Inverse problem of elasticity in other words Cauchy problem (Cauchy-Navier equations of elasticity) has been defined on the accessible outer boundary of the structure. The Cauchy stress tensor components were related with the infinitesimal (incremental calculations) strain tensor components which have been identified in deformed configuration with successive iterations.

The stress-strain relationship in terms of indicial notation is given below:

$$\sigma_{ij} = 2\,\mu\varepsilon_{ij} + \lambda\delta_{ij}\varepsilon_{kk} \tag{1}$$

Here, μ, λ are the Lamé constants. The Cauchy strain components represent the geometrical nonlinearity of the material according to the deformed configuration.

The inverse problem solution depends on the stepwise calculated and so updated Cauchy stress and strain distributions, over the whole boundary of the geometry. Experimentally, tractions and displacements have been measured by nondestructive tests. In isotropic, fiber, and particulate composite material concepts, the stress-strain distributions $\sigma - \varepsilon$ have been examined according to the defined total number of elastic constants in stiffness [C] matrix. The inverse of the stiffness matrix named as the compliance matrix $[S] = [C]^{-1}$ includes the elastic constants in $\varepsilon - \sigma$ strain versus stress equations. In the generalized Hook's law, anisotropic crystalline materials have been defined with 36 constants. Strain energy function has to be used to show that the number of independent material constants can be reduced from 36 to 21. The solution techniques as iterative methods, inverse method, semi-inverse method, variational formulation, finite element method, finite volume method, and meshless method have been listed in **Figure 2**. The experimental solution techniques have been explained by tensile, compression, torsion, impact, and bending mechanical tests. Nondestructive tests (NDT) have been used to obtain informational data from the surfaces of the materials (nanoindentation-hardness testing).

5. Conclusion

In this introduction chapter, the historical development of the elasticity concept and its engineering properties were presented briefly. According to Newton's action and reaction principle, the materials behave linear or nonlinear elastically under typical loading. Elasticity theory provides necessarily required equations and solution techniques. The action-response principle defined between the work done by the forces and the potential energy stored has been explained by the material elastic constants. The mechanical response of a homogeneous isotropic linearly elastic material can be explained by two physical constants, Young's modulus and Poisson's ratio. The elastic properties of particle composites, consisting in a dispersion of nonlinear (spherical or cylindrical) nonhomogeneities into a linear solid matrix, were explained by homogenization procedure. The linear elastic constants of fiber composite materials have been defined according to their three principle directions [6]. These principle directions coincided with the fiber orientations located in each layer. By contrast, the physical-mechanical properties of nonlinear elastic materials have generally been described by parameters which have formations as the scalar functions of the deformation, and their material properties have been determined by selecting the suitable solution techniques.

References

[1] Todhunter I. A history of the theory of elasticity and of strength of the materials from Galilei to the present time, Vol. II. Saint-Venant to Lord Kelvin. King's College, Cambridge: Pearson; 1893. 574 p. http://www.archive.org/details/historyoftheoryo02todhuoft

[2] Todhunter I. A history of the theory of elasticity and of strength of the materials from Galilei to the present time. In: Saint-Venant to Lord Kelvin, Part 1. Vol. 2. King's College Cambridge Pearson; 2014. 348 p. DOI: 10.1017/CB09781107280076

[3] Timoshenko SP, Goodier JN. Theory of Elasticity. 3rd ed. New York: McGraw-Hill; 1970. 580 p

[4] Ugural AC, Fenster S. Advanced Mechanics of Materials and Applied Elasticity. 5th ed. New Jersey: Prentice Hall; 2012. 680 p

[5] Karageorghis A, Lesnic D, Marin L. The method of fundamental solutions for three-dimensional inverse geometric elasticity problems. Computers and Structures. 2016;**166**:51-59. DOI: 10.1016/j.compstruc.2016.01.010

[6] Jones RM. Mechanics of Composite Materials. New York: Hemisphere Publishing Co; 1975. 355 p

Chapter 2

An Overview of Stress-Strain Analysis for Elasticity Equations

Abstract

The present chapter contains the analysis of stress, analysis of strain and stress-strain relationship through particular sections. The theory of elasticity contains equilibrium equations relating to stresses, kinematic equations relating to the strains and displacements and the constitutive equations relating to the stresses and strains. Concept of normal and shear stresses, principal stress, plane stress, Mohr's circle, stress invariants and stress equilibrium relations are discussed in analysis of stress section while strain-displacement relationship for normal and shear strain, compatibility of strains are discussed in analysis of strain section through geometrical representations. Linear elasticity, generalized Hooke's law and stress-strain relations for triclinic, monoclinic, orthotropic, transversely isotropic, fiber-reinforced and isotropic materials with some important relations for elasticity are discussed.

Keywords: analysis of stress, analysis of strain, Mohr's circle, compatibility of strain, stress-strain relation, generalized Hooke's law

1. Introduction

If the external forces producing deformation do not exceed a certain limit, the deformation disappears with the removal of the forces. Thus the elastic behavior implies the absence of any permanent deformation. Every engineering material/composite possesses a certain extent of elasticity. The common materials of construction would remain elastic only for very small strains before exhibiting either plastic straining or brittle failure. However, natural polymeric composites show elasticity over a wider range and the widespread use of natural rubber and similar composites motivated the development of finite elasticity. The mathematical theory of elasticity is possessed with an endeavor to decrease the computation for condition of strain, or relative displacement inside a solid body which is liable to the activity of an equilibrating arrangement of forces, or is in a condition of little inward relative motion and with tries to obtain results which might have been basically essential applications to design, building, and all other helpful expressions in which the material of development is solid.

The elastic properties of continuous materials are determined by the underlying molecular structure, but the relation between material properties and the molecular structure and arrangement in materials is complicated. There are wide classes of materials that might be portrayed by a couple of material constants which can be determined by macroscopic experiments. The quantity of such constants relies upon the nature of the crystalline structure of the material. In this section, we give a short but then entire composition of the basic highlights of applied elasticity having pertinence to our topics. This praiseworthy theory, likely the most successful and best surely understood theory of elasticity, has been given numerous excellent and comprehensive compositions. Among the textbooks including an ample coverage of the problems, we deal with in this chapter which are discussed earlier by Love [1], Sokolnikoff [2], Malvern [3], Gladwell [4], Gurtin [5], Brillouin [6], Pujol [7], Ewing, Jardetsky and Press [8], Achenbach [9], Eringen and Suhubi [10], Jeffreys and Jeffreys [11], Capriz and Podio-Guidugli [12], Truesdell and Noll [13] whose use of direct notation and we find appropriate to avoid encumbering conceptual developments with component-wise expressions. Meriam and Kraige [14] gave an overview of engineering mechanics in theirs book and Podio-Guidugli [15, 16] discussed the strain and examples of concentrated contact interactions in simple bodies in the primer of elasticity. Interestingly, no matter how early in the history of elasticity the consequences of concentrated loads were studied, some of those went overlooked until recently [17–22]. The problem of the determination of stress and strain fields in the elastic solids are discussed by many researchers [23–33]. Belfield et al. [34] discussed the stresses in elastic plates reinforced by fibers lying in concentric circles. Biot [35–38] gave the theory for the propagation of elastic waves in an initially stressed and fluid saturated transversely isotropic media. Borcherdt and Brekhovskikh [39–41] studied the propagation of surface waves in viscoelastic layered media. The fundamental study of seismic surface waves due to the theory of linear viscoelasticity and stress-strain relationship is elaborated by some notable researchers [42–46]. The stress intensity factor is computed due to diffraction of plane dilatational waves by a finite crack by Chang [47], magnetoelastic shear waves in an infinite self-reinforced plate by Chattopadhyay and Choudhury [48]. The propagation of edge wave under initial stress is discussed by Das and Dey [49] and existence and uniqueness of edge waves in a generally anisotropic laminated elastic plates by Fu and Brookes [50, 51]. The basic and historical literature about the stress-strain relationship for propagation of elastic waves in kinds of medium is given by some eminent researchers [52–57]. Kaplunov, Pichugin and Rogersion [58–60] have discussed the propagation of extensional edge waves in in semi-infinite isotropic plates, shells and incompressible plates under the influence of initial stresses. The theory of boundary layers in highly anisotropic and/or reinforced elasticity is studied by Hool, Kinne and Spencer [61, 62].

This chapter addresses the analysis of stress, analysis of strain and stress-strain relationship through particular sections. Concept of normal and shear stress, principal stress, plane stress, Mohr's circle, stress invariants and stress equilibrium relations are discussed in analysis of stress section while strain-displacement relationship for normal and shear strain, compatibility of strains are discussed in analysis of strain section through geometrical representations too. Linear elasticity generalized Hooke's law and stress-strain relation for triclinic, monoclinic, orthotropic, transversely isotropic and isotropic materials are discussed and some important relations for elasticity are deliberated.

2. Analysis of stress

A body consists of huge number of grains or molecules. The internal forces act within a body, representing the interaction between the grains or molecules of the body. In general, if a body is in statically equilibrium, then the internal forces are

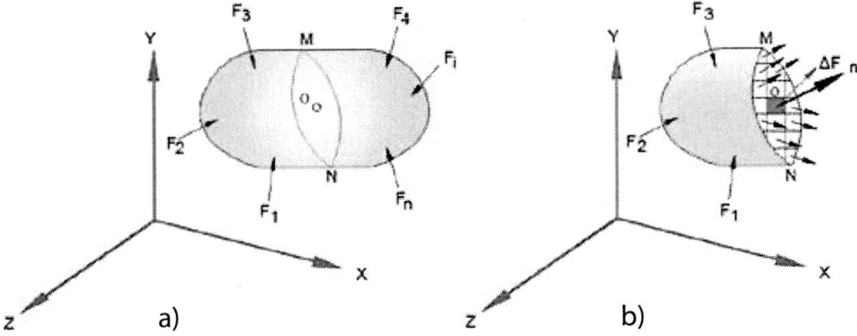

Figure 1.
Forces acting on a (a) body, (b) cross-section of the body.

equilibrated on the basis of Newton's third law. The internal forces are always present even though the external forces are not active.

To examine these internal forces at a point O in **Figure 1(a)**, inside the body, consider a plane MN passing through the point O. If the plane is divided into a number of small areas, as in the **Figure 1(b)**, and the forces acting on each of these are measured, it will be observed that these forces vary from one small area to the next. On the small area ΔA at point O, a force ΔF will be acting as shown in **Figure 1(b)**. From this the concept of stress as the internal force per unit area can be understood. Assuming that the material is continuous, the term "stress" at any point across a small area ΔA can be defined by the limiting equation as below.

$$\textbf{Stress } (\sigma) = \lim_{\Delta A \to 0} \frac{\Delta F}{\Delta A} \tag{1}$$

where ΔF is the internal force on the area ΔA surrounding the given point. Forces which act on an element of material may be of two types:

i. body forces and

ii. surface forces.

Body forces always act on every molecule of a body and are proportional to the volume whereas surface force acts over the surface of the body and is measure in terms of force per unit area. The force acting on a surface may resolve into normal stress and shear stress. Normal stress may be tensile or compressive in nature. Positive side of normal stress is for tensile stress whilst negative side is for compressive.

2.1 Concept of normal stress and shear stress

Figure 2(a) shows the rectangular components of the force vector ΔF referred to corresponding axes. Taking the ratios $\Delta F_x/\Delta A_x$, $\Delta F_y/\Delta A_x$, $\Delta F_z/\Delta A_x$, three quantities that set up the average intensity of the force on the area ΔA_x When the limit $\Delta A \to 0$, the above ratios are characterized as the force intensity acting on X-face at point O. These values associated with three intensities are defined as the "Stress components" related with the X-face at point O. The stress component parallel to the surface are called "Shear stress component," is indicated by τ. The

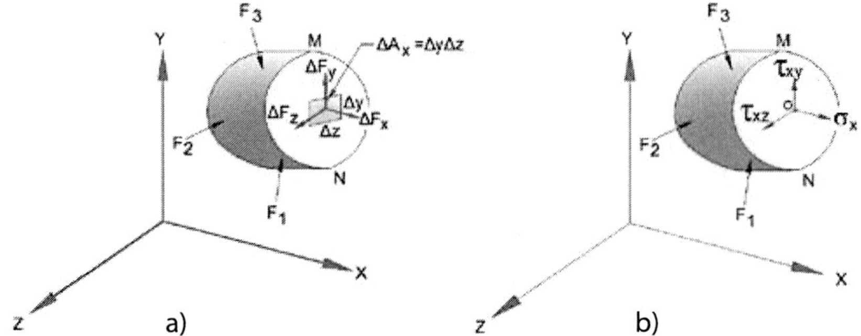

Figure 2.
(a) Force components of ΔF acting on small area centered at point O and (b) stress components at point O.

shear stress component acting on the X-face in the Y-direction is identified as τ_{xy}. The stress component perpendicular to the face is called "Normal Stress" or "Direct stress" component and is denoted by σ.

From the above discussions, the stress components on the X-face at point O are defined as follows in terms of force intensity ratios

$$\left.\begin{aligned}\sigma_x &= \lim_{\Delta A_x \to 0} \frac{\Delta F_x}{\Delta A_x} \\ \tau_{xy} &= \lim_{\Delta A_x \to 0} \frac{\Delta F_y}{\Delta A_x} \\ \tau_{xz} &= \lim_{\Delta A_x \to 0} \frac{\Delta F_z}{\Delta A_x}\end{aligned}\right\} \tag{2}$$

and the above stress components are illustrated in **Figure 2(b)**.

2.2 Stress components

Three mutually perpendicular coordinate axes x, y, z are taken. We consider the stresses act on the surface of the cubic element of the substance. When a force is applied, as mean that the state of stress is perfectly homogeneous throughout the element and that the body is in equilibrium as shown in **Figure 3**. There are nine quantities which are acting on the faces of the cubic and are known as the stress components.

In matrix notation, the stress components can be written as

$$\begin{pmatrix} \sigma_x & \tau_{xy} & \tau_{xz} \\ \tau_{yx} & \sigma_y & \tau_{yz} \\ \tau_{zx} & \tau_{zy} & \sigma_z \end{pmatrix} \tag{3}$$

which completely define the state of stress in the elemental cube. The first suffix of the shear stress refers to the normal to the plane on which the stress acts and the second suffix refer to the direction of shear stress on this plane. The nine stress components which are derived in matrix form are not all independent quantities.

2.3 Principal stress and stress invariants

Let us consider three mutually perpendicular planes in which shear stress is zero and on these planes the normal stresses have maximum or minimum values. These

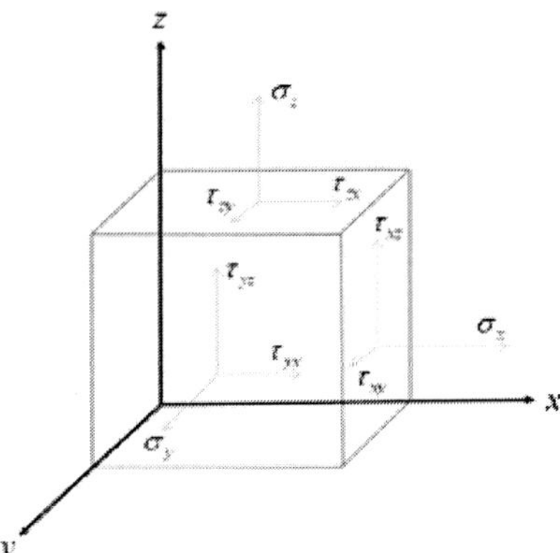

Figure 3.
Stress components acting on cube.

normal stresses are referred to as principal stresses and the plane in which these normal stresses act is called principal plane.

Invariants mean those amounts that are unexchangeable and do not differ under various conditions. With regards to stress components, invariants are such quantities that don't change with rotation of axes or which stay unaffected under transformation, from one set of axes to another. Subsequently, the combination of stresses at a point that don't change with the introduction of co-ordinate axis is called stress invariants.

2.4 Plane stress

Numerous metal shaping procedures include biaxial condition of stress. On the off chance that one of the three normal and shear stresses acting on a body is zero, the state of stress is called plane stress condition. All stresses act parallel to x and y axes. Plane pressure condition is gone over in numerous engineering and forming applications. Regularly, slip can be simple if the shear stress following up on the slip planes is adequately high and acts along favored slip direction. Slip planes may be inclined with respect to the external stress acting on solids. It becomes necessary to transform the stresses acting along the original axes into the inclined planes. Stress change ends up essential in such cases.

2.4.1 Stress transformation in plane stress

Consider the plane stress condition acting on a plane as shown in **Figure 4**. Let us investigate the state of stresses onto a transformed plane which is inclined at an angle θ with respect to x, y axes.

Let by rotating of the x and y axes through the angle θ, a new set of axes X' and Y' will be formed. The stresses acting on the plane along the new axes are obtained when the plane has been rotated about the z axis. In order to obtain these transformed stresses, we take equilibrium of forces on the inclined plane both perpendicular to and parallel to the inclined plane.

Thus, the expression for transformed stress using the direction cosines can be written as

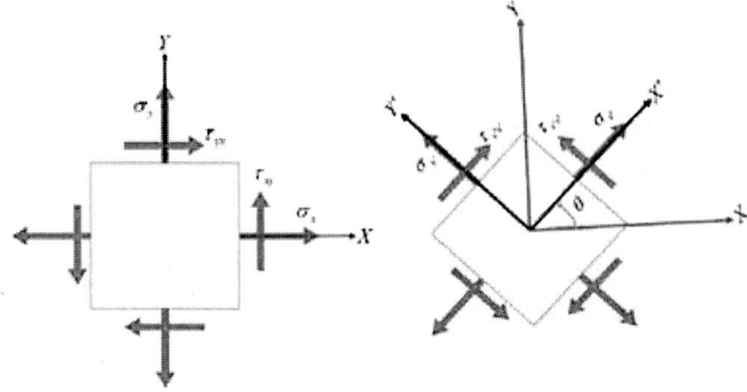

Figure 4.
Representation of stresses on inclined plane.

$$\begin{aligned}\sigma_{x'} &= l_{x'x}^2\sigma_x + l_{x'y}^2\sigma_y + 2l_{x'x}l_{x'y}\tau_{xy} \\ &= 2\cos^2\theta\sigma_x + 2\sin^2\theta\sigma_y + 2\cos\theta\sin\theta\tau_{xy}\end{aligned} \tag{4}$$

Similarly, write for the *y'* normal stress and shear stress.
The transformed stresses are given as

$$\begin{aligned}\sigma_{x'} &= \frac{\sigma_x + \sigma_y}{2} + \frac{\sigma_x - \sigma_y}{2}\cos 2\theta + \tau_{xy}\sin 2\theta \\ \sigma_{y'} &= \frac{\sigma_x + \sigma_y}{2} - \frac{\sigma_x - \sigma_y}{2}\cos 2\theta - \tau_{xy}\sin 2\theta\end{aligned} \tag{5}$$

and

$$\tau_{x'y'} = \frac{\sigma_y - \sigma_x}{2}\sin 2\theta + \tau_{xy}\cos 2\theta$$

where $\sigma_{x'}$ and $\tau_{x'y'}$ are respectively the normal and shear stress acting on the inclined plane. The above three equations are known as transformation equations for plane stress.

In order to design components against failure the maximum and minimum normal and shear stresses acting on the inclined plane must be derived. The maximum normal stress and shear stress can be found when we differentiate the stress transformation equations with respect to θ and equate to zero. The maximum and minimum stresses are known as principal stresses and the plane of acting is named as principal planes.

Maximum normal stress is given by

$$\sigma_1, \sigma_2 = \frac{\sigma_x + \sigma_y}{2} \pm \sqrt{\left(\frac{\sigma_x - \sigma_y}{2}\right)^2 + \tau_{xy}^2} \tag{6}$$

and maximum shear stress is

$$\tau_{\max} = \sqrt{\left(\frac{\sigma_x - \sigma_y}{2}\right)^2 + \tau_{xy}^2} \tag{7}$$

$$\text{with } \tau_{\max} = \frac{\sigma_1 - \sigma_2}{2}. \tag{8}$$

The plane on which the principal normal stress acts, the shear stress is zero and vice versa. The angle corresponding to the principal planes can be obtained from $\tan 2\theta = \frac{\tau_{xy}}{\frac{\sigma_x - \sigma_y}{2}}$ for the principal normal planes and $\tan 2\theta = \frac{\tau_{xy}}{\frac{\sigma_x - \sigma_y}{2}}$ is for the principal shear plane.

2.4.2 Mohr's circle for plane stress

The transformation equations of plane stress which are given by Eq. (5) can be represented in a graphical form (**Figure 5**) by *Mohr's circle*. The transformation equations are sufficient to get the normal and shear stresses on any plane at a point, with Mohr's circle one can easily visualize their variation with respect to plane orientation θ.

2.4.2.1 Equations of Mohr's circle

Rearranging the terms of Eq. (5), we get

$$\sigma_{x'} - \frac{\sigma_x + \sigma_y}{2} = \frac{\sigma_x - \sigma_y}{2} \cos 2\theta + \tau_{xy} \sin 2\theta \tag{9.1}$$

and

$$\tau_{x'y'} = -\left(\frac{\sigma_x - \sigma_y}{2}\right) \sin 2\theta + \tau_{xy} \cos 2\theta \tag{9.2}$$

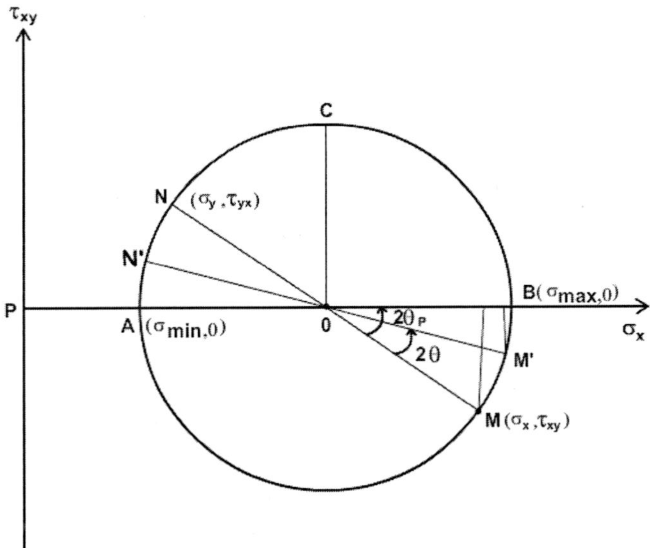

Figure 5.
Mohr's circle diagram.

Squaring and adding the Eqs. (9.1) and (9.2), result in

$$\left(\sigma_{x'} - \frac{\sigma_x + \sigma_y}{2}\right)^2 + \tau_{x'y'}^2 = \left(\frac{\sigma_x - \sigma_y}{2}\right)^2 + \tau_{xy}^2 \tag{10}$$

For simple representation of Eq. (10), the following notations are used

$$\sigma_{av} = \frac{\sigma_x + \sigma_y}{2}, \; r = \left[\left(\frac{\sigma_x - \sigma_y}{2}\right)^2 + \tau_{xy}^2\right]^{1/2} \tag{11}$$

Thus, the simplified form of Eq. (10) can be written as

$$(\sigma_{x'} - \sigma_{av})^2 + \tau_{x'y'}^2 = r^2 \tag{12}$$

Eq. (12) represents the equation of a circle in a standard form. This circle has $\sigma_{x'}$ as its abscissa and $\tau_{x'y'}$ as its ordinate with radius r. The coordinate for the center of the circle is $(\sigma_{av}, 0)$.

Mohr's circle is drawn by considering the stress coordinates σ_x as its abscissa and τ_{xy} as its ordinate, and this plane is known as the stress plane. The plane on the element bounded with xy coordinates in the material is named as physical plane. Stresses on the physical plane M is represented by the point M on the stress plane with σ_x and τ_{xy} coordinates.

Stresses on the physical plane which is normal to i.e. N, is given by the point N on the stress plane with σ_y and τ_{yx}. O is the intersecting point of line MN and which is at the center of the circle and radius of the circle is OM. Now, the stresses on a plane, making θ inclination with x axis in physical plane can be determined as follows.

An important point to be noted here is that a plane which has a θ inclination in physical plane will make 2θ inclination in stress plane M. Hence, rotate the line OM in stress plane by 2θ counter clockwise to obtain the plane M'. The coordinates of M' in stress plane define the stresses acting on plane M' in physical plane and it can be easily verified.

$$\sigma_{x'} = PO + r\cos\left(2\theta_p - 2\theta\right) \tag{13}$$

where $PO = \frac{\sigma_x + \sigma_y}{2}$, $r = \left[\left(\frac{\sigma_x - \sigma_y}{2}\right)^2 + \tau_{xy}^2\right]^{1/2}$, $\cos 2\theta_p = \frac{\sigma_x - \sigma_y}{2r}$, $\sin 2\theta_p = \frac{\tau_{xy}}{2r}$.
On simplifying Eq. (13)

$$\sigma_{x'} = \frac{\sigma_x + \sigma_y}{2} + \frac{\sigma_x - \sigma_y}{2}\cos 2\theta + \tau_{xy}\sin 2\theta \tag{14}$$

Eq. (14) is same as the first equation of Eq. (5).
This way it can be proved for shear stress $\tau_{x'y'}$ on plane M' (do yourself).

2.4.3 Stress equilibrium relation

Let σ_x, τ_{yx}, τ_{zx} are the stress components acting along the x-direction, τ_{xy}, σ_y, τ_{zy} are the stress components acting along the y-direction and τ_{xz}, τ_{yz}, σ_z are the stress components acting along the z-direction. The body forces F_x, F_y, F_z acting along x,

y, z direction respectively. Then the stress equilibrium relation or equation of motion in terms of stress components are given by

$$\left.\begin{aligned}\frac{\partial\sigma_x}{\partial x}+\frac{\partial\tau_{yx}}{\partial y}+\frac{\partial\tau_{zx}}{\partial z}+F_x=0,\\\frac{\partial\tau_{xy}}{\partial x}+\frac{\partial\sigma_y}{\partial y}+\frac{\partial\tau_{zy}}{\partial z}+F_y=0,\\\frac{\partial\tau_{xz}}{\partial x}+\frac{\partial\tau_{yz}}{\partial y}+\frac{\partial\sigma_z}{\partial z}+F_z=0.\end{aligned}\right\}\tag{15}$$

3. Analysis of strain

While defining a stress it was pointed out that stress is an abstract quantity which cannot be seen and is generally measured indirectly. Strain differs in this respect from stress. It is a complete quantity that can be seen and generally measured directly as a relative change of length or shape. In generally, stress is the ratio of change in original dimension and the original dimension. It is the dimensionless constant quantity.

3.1 Types of strain

Strain may be classified into three types; normal strain, shear strain and volumetric strain.

The normal strain is the relative change in length whether shearing strain relative change in shape. The volumetric strain is defined by the relative change in volume.

3.2 Strain-displacement relationship

3.2.1 Normal strain

Consider a line element of length Δx emanating from position (x, y) and lying in the x-direction, denoted by AB in **Figure 6**. After deformation the line element occupies $A'B'$, having undergone a translation, extension and rotation.

The particle that was originally at x has undergone a displacement $u_x(x,y)$ and the other end of the line element has undergone a displacement $u_x(x+\Delta x,y)$. By the definition of normal strain

$$\varepsilon_{xx}=\frac{A'B^{*}-AB}{AB}=\frac{u_x(x+\Delta x,y)-u_x(x,y)}{\Delta x}.\tag{16}$$

In the limit $\Delta x\to 0$, Eq. (16) becomes

$$\varepsilon_{xx}=\frac{\partial u_x}{\partial x}\tag{17}$$

This partial derivative is a **displacement gradient**, a measure of how rapid the displacement changes through the material, and is the strain at (x, y). Physically, it represents the (approximate) unit change in length of a line element.

Similarly, by considering a line element initially lying in the y-direction, the strain in the y-direction can be expressed as

$$\varepsilon_{yy}=\frac{\partial u_y}{\partial y}.\tag{18}$$

3.2.2 Shear strain

The particles A and B in **Figure 6** also undergo displacements in the y-direction and this is shown in **Figure 7(a)**. In this case, we have

$$B^*B' = \frac{\partial u_y}{\partial x}\Delta x. \tag{19}$$

A similar relation can be derived by considering a line element initially lying in the y-direction. From the **Figure 7(b)**, we have

$$\theta \approx \tan\theta = \frac{\partial u_y/\partial x}{1+\partial u_x/\partial x} \approx \frac{\partial u_y}{\partial x} \tag{20}$$

provided that (i) θ is small and (ii) the displacement gradient $\partial u_x/\partial x$ is small. A similar expression for the angle λ can be derived as

$$\lambda \approx \frac{\partial u_x}{\partial y} \tag{21}$$

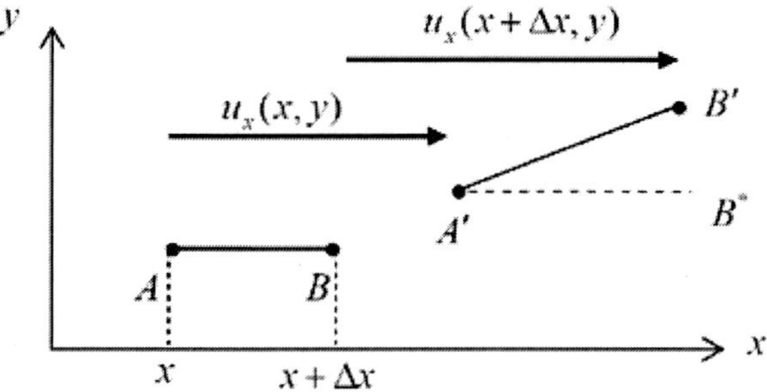

Figure 6.
Deformation of a line element.

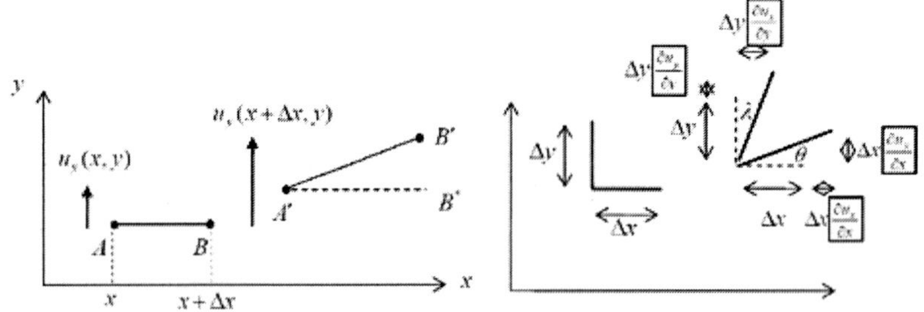

Figure 7.
(a) Deformation of a line element and (b) strains in terms of displacement gradients.

and hence the shear strain can be written in terms of displacement gradients as

$$\varepsilon_{xy} = \frac{1}{2}\left(\frac{\partial u_x}{\partial y} + \frac{\partial u_y}{\partial x}\right). \tag{22}$$

In similar manner, the strain-displacement relation for three dimensional body is given by

$$\begin{aligned}
&\varepsilon_{xx} = \frac{\partial u_x}{\partial x},\ \varepsilon_{yy} = \frac{\partial u_y}{\partial y},\ \varepsilon_{zz} = \frac{\partial u_z}{\partial z},\\
&\varepsilon_{xy} = \frac{1}{2}\left(\frac{\partial u_x}{\partial y} + \frac{\partial u_y}{\partial x}\right),\ \varepsilon_{xz} = \frac{1}{2}\left(\frac{\partial u_x}{\partial z} + \frac{\partial u_z}{\partial x}\right),\ \varepsilon_{yz} = \frac{1}{2}\left(\frac{\partial u_y}{\partial z} + \frac{\partial u_z}{\partial y}\right).
\end{aligned} \tag{23}$$

3.3 Compatibility of strain

As seen in the previous section, there are three strain-displacement relations Eqs. (17), (18) and (22) but only two displacement components. This implies that the strains are not independent but are related in some way. The relations between the strains are called compatibility conditions.

3.3.1 Compatibility relations

Let us suppose that the point P which is act (x,y) before straining and it will be at P' after straining on the co-ordinate plane Oxy as depicted in **Figure 8**. Then (u,v) is a displacement corresponding to the point P. The variable u and v are the functions of x and y.

Using the fundamental notation

$$\varepsilon_{xx} = \frac{\partial u_x}{\partial x},\ \varepsilon_{yy} = \frac{\partial u_y}{\partial y},\ \varepsilon_{xy} = \frac{1}{2}\left(\frac{\partial u_x}{\partial y} + \frac{\partial u_y}{\partial x}\right) \tag{24}$$

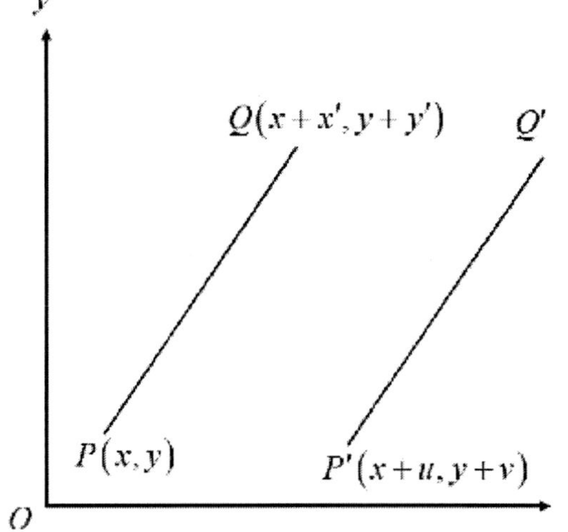

Figure 8.
Deformation of line element.

we get

$$\frac{\partial^2 \varepsilon_{xx}}{\partial y^2} = \frac{\partial^3 u_x}{\partial x \partial y^2}, \frac{\partial^2 \varepsilon_{yy}}{\partial x^2} = \frac{\partial^3 u_y}{\partial x^2 \partial y} \tag{25}$$

$$\frac{\partial^2 \varepsilon_{xy}}{\partial x \partial y} = \frac{1}{2}\left(\frac{\partial^3 u_y}{\partial x^2 \partial y} + \frac{\partial^3 u_x}{\partial x \partial y^2}\right). \tag{26}$$

Eqs. (25) and (26) result in

$$\frac{\partial^2 \varepsilon_{xy}}{\partial x \partial y} = \frac{1}{2}\left(\frac{\partial^2 \varepsilon_{xx}}{\partial y^2} + \frac{\partial^2 \varepsilon_{yy}}{\partial x^2}\right) \tag{27}$$

which is the compatibility condition in two dimension.

4. Stress-strain relation

In the previous section, the state of stress at a point was characterized by six components of stress, and the internal stresses and the applied forces are accompanied with the three equilibrium equation. These equations are applicable to all types of materials as the relationships are independent of the deformations (strains) and the material behavior.

Also, the state of strain at a point was defined in terms of six components of strain. The strains and the displacements are related uniquely by the derivation of six strain-displacement relations and compatibility equations. These equations are also applicable to all materials as they are independent of the stresses and the material behavior and hence.

Irrespective of the independent nature of the equilibrium equations and strain-displacement relations, usually, it is essential to study the general behavior of materials under applied loads including these relations. Strains will be developed in a body due to the application of a load, stresses and deformations and hence it is become necessary to study the behavior of different types of materials. In a general three-dimensional system, there will be 15 unknowns namely 3 displacements, 6 strains and 6 stresses. But we have only 9 equations such as 3 equilibrium equations and 6 strain-displacement equations to achieve these 15 unknowns. It is important to note that the compatibility conditions are not useful for the determination of either the displacements or strains. Hence the additional six equations relating six stresses and six strains will be developed. These equations are known as "Constitutive equations" because they describe the macroscopic behavior of a material based on its internal constitution.

4.1 Linear elasticity generalized Hooke's law

Hooke's law provides the unique relationship between stress and strain, which is independent of time and loading history. The law can be used to predict the deformations used in a given material by a combination of stresses.

The linear relationship between stress and strain is given by

$$\sigma_x = E\varepsilon_{xx} \tag{28}$$

where E is known as Young's modulus.

In general, each strain is dependent on each stress. For example, the strain ε_{xx} written as a function of each stress as

$$\varepsilon_{xx} = C_{11}\sigma_x + C_{12}\sigma_y + C_{13}\sigma_z + C_{14}\tau_{xy} + C_{15}\tau_{yz} + C_{16}\tau_{zx} + C_{17}\tau_{xz} + C_{18}\tau_{zy} + C_{19}\tau_{yx}. \quad (29)$$

Similarly, stresses can be expressed in terms of strains which state that at each point in a material, each stress component is linearly related to all the strain components. This is known as **generalized Hook's law.**

For the most general case of three-dimensional state of stress, Eq. (28) can be written as

$$\left(\sigma_{ij}\right)_{9\times 1} = \left(D_{ijkl}\right)_{9\times 9}\left(\varepsilon_{kl}\right)_{9\times 1} \quad (30)$$

where $\left(D_{ijkl}\right)$ is elasticity matrix, $\left(\sigma_{ij}\right)$ is stress components, $\left(\varepsilon_{kl}\right)$ is strain components.

Since both stress σ_{ij} and strain ε_{ij} are second-order tensors, it follows that D_{ijkl} is a fourth order tensor, which consists of $3^4 = 81$ material constants if symmetry is not assumed.

Now, from $\sigma_{ij} = \sigma_{ji}$ and $\varepsilon_{ij} = \varepsilon_{ji}$, the number of 81 material constants is reduced to 36 under symmetric conditions of $D_{ijkl} = D_{jikl} = D_{ijlk} = D_{jilk}$ which provides stress-strain relation for most general form of anisotropic material.

4.1.1 Stress-strain relation for triclinic material

The stress-strain relation for triclinic material will consist 21 elastic constants which is given by

$$\begin{bmatrix} \sigma_x \\ \sigma_y \\ \sigma_z \\ \tau_{xy} \\ \tau_{yz} \\ \tau_{zx} \end{bmatrix} = \begin{bmatrix} D_{11} & D_{12} & D_{13} & D_{14} & D_{15} & D_{16} \\ D_{12} & D_{22} & D_{23} & D_{24} & D_{25} & D_{26} \\ D_{13} & D_{23} & D_{33} & D_{34} & D_{35} & D_{36} \\ D_{14} & D_{24} & D_{34} & D_{44} & D_{45} & D_{46} \\ D_{15} & D_{25} & D_{35} & D_{45} & D_{55} & D_{56} \\ D_{16} & D_{26} & D_{36} & D_{46} & D_{56} & D_{66} \end{bmatrix} \begin{bmatrix} \varepsilon_{xx} \\ \varepsilon_{yy} \\ \varepsilon_{zz} \\ \varepsilon_{xy} \\ \varepsilon_{yz} \\ \varepsilon_{zx} \end{bmatrix}. \quad (31)$$

4.1.2 Stress-strain relation for monoclinic material

The stress-strain relation for monoclinic material will consist 13 elastic constants which is given by

$$\begin{bmatrix} \sigma_x \\ \sigma_y \\ \sigma_z \\ \tau_{xy} \\ \tau_{yz} \\ \tau_{zx} \end{bmatrix} = \begin{bmatrix} D_{11} & D_{12} & D_{13} & 0 & D_{15} & 0 \\ D_{12} & D_{22} & D_{23} & 0 & D_{25} & 0 \\ D_{13} & D_{23} & D_{33} & 0 & D_{35} & 0 \\ 0 & 0 & 0 & D_{44} & 0 & D_{46} \\ D_{15} & D_{25} & D_{35} & 0 & D_{55} & 0 \\ 0 & 0 & 0 & D_{46} & 0 & D_{66} \end{bmatrix} \begin{bmatrix} \varepsilon_{xx} \\ \varepsilon_{yy} \\ \varepsilon_{zz} \\ \varepsilon_{xy} \\ \varepsilon_{yz} \\ \varepsilon_{zx} \end{bmatrix}. \quad (32)$$

4.1.3 Stress-strain relation for orthotropic material

A material that exhibits symmetry with respect to three mutually orthogonal planes is called an orthotropic material. The stress-strain relation for orthotropic material will consist 9 elastic constants which is given by

$$\begin{bmatrix} \sigma_x \\ \sigma_y \\ \sigma_z \\ \tau_{xy} \\ \tau_{yz} \\ \tau_{zx} \end{bmatrix} = \begin{bmatrix} D_{11} & D_{12} & D_{13} & 0 & 0 & 0 \\ D_{12} & D_{22} & D_{23} & 0 & 0 & 0 \\ D_{13} & D_{23} & D_{33} & 0 & 0 & 0 \\ 0 & 0 & 0 & D_{44} & 0 & 0 \\ 0 & 0 & 0 & 0 & D_{55} & 0 \\ 0 & 0 & 0 & 0 & 0 & D_{66} \end{bmatrix} \begin{bmatrix} \varepsilon_{xx} \\ \varepsilon_{yy} \\ \varepsilon_{zz} \\ \varepsilon_{xy} \\ \varepsilon_{yz} \\ \varepsilon_{zx} \end{bmatrix}. \tag{33}$$

4.1.4 Stress-strain relation for transversely isotropic material

Transversely isotropic material exhibits a rationally elastic symmetry about one of the coordinate axes x, y and z. In such case, the material constants reduce to 5 as shown below

$$\begin{bmatrix} \sigma_x \\ \sigma_y \\ \sigma_z \\ \tau_{xy} \\ \tau_{yz} \\ \tau_{zx} \end{bmatrix} = \begin{bmatrix} D_{11} & D_{12} & D_{13} & 0 & 0 & 0 \\ D_{12} & D_{11} & D_{13} & 0 & 0 & 0 \\ D_{13} & D_{13} & D_{33} & 0 & 0 & 0 \\ 0 & 0 & 0 & (D_{11}-D_{12})/2 & 0 & 0 \\ 0 & 0 & 0 & 0 & D_{44} & 0 \\ 0 & 0 & 0 & 0 & 0 & D_{44} \end{bmatrix} \begin{bmatrix} \varepsilon_{xx} \\ \varepsilon_{yy} \\ \varepsilon_{zz} \\ \varepsilon_{xy} \\ \varepsilon_{yz} \\ \varepsilon_{zx} \end{bmatrix}. \tag{34}$$

4.1.5 Stress-strain relation for fiber-reinforced material

The constitutive equation for a fiber-reinforced material whose preferred direction is that of a unit vector $\vec{a}$ is

$$\begin{aligned} \tau_{ij} = \lambda e_{kk}\delta_{ij} + 2\mu_T e_{ij} + \alpha\left(a_k a_m e_{km}\delta_{ij} + e_{kk}a_i a_j\right) + 2(\mu_L - \mu_T)\left(a_i a_k e_{kj} + a_j a_k e_{ki}\right) \\ + \beta a_k a_m e_{km} a_i a_j; \quad i, j, k, m = 1, 2, 3 \end{aligned} \tag{35}$$

where τ_{ij} are components of stress, e_{ij} are components of infinitesimal strain, and a_i the components of $\vec{a}$, which are referred to rectangular Cartesian co-ordinates x_i. The vector $\vec{a}$ may be a function of position. Indices take the value 1, 2 and 3, and the repeated suffix summation convention is adopted. The coefficients λ, μ_L, μ_T, α and β are all elastic constant with the dimension of stress.

4.1.6 Stress-strain relation for isotropic material

For a material whose elastic properties are not a function of direction at all, only two independent elastic material constants are sufficient to describe its behavior completely. This material is called isotropic linear elastic. The stress-strain relationship for this material is written as

$$\begin{bmatrix} \sigma_x \\ \sigma_y \\ \sigma_z \\ \tau_{xy} \\ \tau_{yz} \\ \tau_{zx} \end{bmatrix} = \begin{bmatrix} D_{11} & D_{12} & D_{12} & 0 & 0 & 0 \\ D_{12} & D_{11} & D_{12} & 0 & 0 & 0 \\ D_{12} & D_{12} & D_{11} & 0 & 0 & 0 \\ 0 & 0 & 0 & (D_{11}-D_{12})/2 & 0 & 0 \\ 0 & 0 & 0 & 0 & (D_{11}-D_{12})/2 & 0 \\ 0 & 0 & 0 & 0 & 0 & (D_{11}-D_{12})/2 \end{bmatrix} \begin{bmatrix} \varepsilon_{xx} \\ \varepsilon_{yy} \\ \varepsilon_{zz} \\ \varepsilon_{xy} \\ \varepsilon_{yz} \\ \varepsilon_{zx} \end{bmatrix} \tag{36}$$

which consists only two independent elastic constants. Replacing D_{12} and $D_{12}\,(D_{11} - D_{12})/2$ by λ and μ which are called Lame's constants and in particular μ is also called shear modulus of elasticity, we get

$$\left.\begin{aligned} \sigma_x &= (2\mu + \lambda)\varepsilon_{xx} + \lambda(\varepsilon_{yy} + \varepsilon_{zz}), \\ \sigma_y &= (2\mu + \lambda)\varepsilon_{yy} + \lambda(\varepsilon_{xx} + \varepsilon_{zz}), \\ \sigma_z &= (2\mu + \lambda)\varepsilon_{zz} + \lambda(\varepsilon_{yy} + \varepsilon_{xx}), \\ \tau_{xy} &= \mu\varepsilon_{xy}, \tau_{yz} = \mu\varepsilon_{yz}, \tau_{zx} = \mu\varepsilon_{zx}. \end{aligned}\right\} \tag{37}$$

Also, from the above relation some important terms are induced which are as follow

(1) **Bulk modulus:** Bulk modulus is the relative change in the volume of a body produced by a unit compressive or tensile stress acting uniformly over its surface. Symbolically

$$K = \lambda + \frac{2}{3}\mu. \tag{38}$$

(2) **Young's modulus:** Young's modulus is a measure of the ability of a material to withstand changes in length when under lengthwise tension or compression. Symbolically

$$E = \frac{\mu(3\lambda + 2\mu)}{\lambda + \mu}. \tag{39}$$

(3) **Poisson's ratio:** The ratio of transverse strain and longitudinal strain is called Poisson's ratio. Symbolically

$$\nu = \frac{\lambda}{2(\lambda + \mu)}. \tag{40}$$

5. Conclusions

This chapter dealt the analysis of stress, analysis of strain and stress-strain relationship through particular sections. Concept of normal and shear stress, principal stress, plane stress, Mohr's circle, stress invariants and stress equilibrium relations are discussed in analysis of stress section while strain-displacement relationship for normal and shear strain, compatibility of strains are discussed in analysis of strain section through geometrical representations. Linear elasticity, generalized Hooke's law and stress-strain relation for triclinic, monoclinic, orthotropic, transversely-isotropic, fiber-reinforced and isotropic materials with some important relations for elasticity are discussed mathematically.

References

[1] Love AEH. A Treatise on the Mathematical Theory of Elasticity. Cambridge: Cambridge University Press; 1927

[2] Sokolnikoff IS. Mathematical Theory of Elasticity. New York: McGraw-Hill; 1956

[3] Malvern LE. Introduction to the Mechanics of a Continuous Medium. Englewood Cliffs: Prentice-Hall; 1969

[4] Gladwell GM. Contact Problems in the Classical Theory of Elasticity. Netherland: Springer Science & Business Media; 1980

[5] Gurtin ME. The linear theory of elasticity. In: Linear Theories of Elasticity and Thermoelasticity. Berlin, Heidelberg: Springer; 1973. pp. 1-295

[6] Brillouin L. Tensors in Mechanics and Elasticity. New York: Academic; 1964

[7] Pujol J. Elastic Wave Propagation and Generation in Seismology. Cambridge, New York: Cambridge University Press; 2003

[8] Ewing WM, Jardetsky WS, Press F. Elastic Waves in Layered Media. New York: McGraw-Hill; 1957

[9] Achenbach J. Wave Propagation in Elastic Solids. Amsterdam: North-Holland; 1973

[10] Eringen A, Suhubi E. Elastodynamics. Vol. II. New York: Academic; 1975

[11] Jeffreys H, Jeffreys B. Methods of Mathematical Physics. Cambridge: Cambridge University Press; 1956

[12] Capriz G, Podio-Guidugli P. Whence the boundary conditions in modern continuum physics? In: Capriz G, Podio-Guidugli P, editors. Proceedings of the Symposium. Rome: Accademia dei Lincei; 2004

[13] Truesdell C, Noll W. The non-linear field theories of mechanics. In: The Non-Linear Field Theories of Mechanics. Berlin, Heidelberg: Springer; 2004. pp. 1-579

[14] Meriam JL, Kraige LG. Engineering Mechanics. Vol. 1. New York: Wiley; 2003

[15] Podio-Guidugli P. Strain. In: A Primer in Elasticity. Dordrecht: Springer; 2000

[16] Podio-Guidugli P. Examples of concentrated contact interactions in simple bodies. Journal of Elasticity. 2004;**75**:167-186

[17] Schuricht F. A new mathematical foundation for contact interactions in continuum physics. Archive for Rational Mechanics and Analysis. 2007;**184**: 495-551

[18] Aki K, Richards PG. Quantitative Seismology. Vol. 2. San Francisco: W. H. Freeman & Co; 1980

[19] Mase G. Theory and Problems of Continuum Mechanics, Schaum's outline series. New York: McGraw-Hill; 1970

[20] Morse P, Feshbach H. Methods of Theoretical Physics. Vol. 2. New York: McGraw-Hill; 1953

[21] Mahanty M, Chattopadhyay A, Dhua S, Chatterjee M. Propagation of shear waves in homogeneous and inhomogeneous fibre-reinforced media on a cylindrical Earth model. Applied Mathematical Modelling. 2017;**52**: 493-511

[22] Kumar P, Chattopadhyay A, Singh AK. Shear wave propagation due to a point source. Procedia Engineering. 2017;**173**:1544-1551

[23] Chattopadhyay A, Singh P, Kumar P, Singh AK. Study of Love-type wave propagation in an isotropic tri layers elastic medium overlying a semi-infinite elastic medium structure. Waves in Random and Complex Media. 2017;**28** (4):643-669

[24] Chattopadhyay A, Saha S, Chakraborty M. Reflection and transmission of shear waves in monoclinic media. International Journal for Numerical and Analytical Methods in Geomechanics. 1997;**21**(7):495-504

[25] Chattopadhyay A. Wave reflection and refraction in triclinic crystalline media. Archive of Applied Mechanics. 2004;**73**(8):568-579

[26] Chattopadhyay A, Singh AK. Propagation of magnetoelastic shear waves in an irregular self-reinforced layer. Journal of Engineering Mathematics. 2012;**75**(1):139-155

[27] Udias A, Buforn E. Principles of Seismology. Cambridge, New York: Cambridge University Press; 2017

[28] Novotny O. Seismic Surface Waves. Bahia, Salvador: Instituto de Geociencias; 1999

[29] Brillouin L. Wave Propagation and Group Velocity. Academic Press, New York; 1960

[30] Badriev IB, Banderov VV, Makarov MV, Paimushin VN. Determination of stress-strain state of geometrically nonlinear sandwich plate. Applied Mathematical Sciences. 2015;**9**(77–80): 3887-3895

[31] Jaeger JC, Cook NG, Zimmerman R. Fundamentals of Rock Mechanics. Malden, U.S.A: Blackwell Publishing. 2009

[32] Press F. Seismic wave attenuation in the crust. Journal of Geophysical Research. 1964;**69**(20):4417-4418

[33] Ben-Menahem A, Singh SJ. Seismic Waves and Sources. New York: Springer Verlag; 1981

[34] Belfield AJ, Rogers TG, Spencer AJM. Stress in elastic plates reinforced by fibres lying in concentric circles. Journal of the Mechanics and Physics of Solids. 1983;**31**(1):25-54

[35] Biot MA. The influence of initial stress on elastic waves. Journal of Applied Physics. 1940;**11**(8):522-530

[36] Biot MA. Theory of propagation of elastic waves in a fluid-saturated porous solid. I. Low frequency range. The Journal of the Acoustical Society of America. 1956;**28**(2):168-178

[37] Biot MA. Theory of propagation of elastic waves in a fluid-saturated porous solid. II. Higher frequency range. The Journal of the Acoustical Society of America. 1956;**28**(2):179-191

[38] Biot MA, Drucker DC. Mechanics of incremental deformation. Journal of Applied Mechanics. 1965;**32**(1):957.

[39] Borcherdt RD. Rayleigh-type surface wave on a linear viscoelastic half-space. The Journal of the Acoustical Society of America. 1974;**55**(1):13-15

[40] Borcherdt RD. Viscoelastic Waves in Layered Media. Cambridge, New York: Cambridge University Press; 2009

[41] Brekhovskikh L. Waves in Layered Media. New York: Academic Press; 1980

[42] Bullen KE. Compressibility-pressure hypothesis and the Earth's interior. Geophysical Journal International. 1949; 5:335-368

[43] Bullen KE, Bullen KE, Bolt BA. An Introduction to the Theory of Seismology. Cambridge, New York: Cambridge University Press; 1985

[44] Chapman C. Fundamentals of Seismic Wave Propagation. Cambridge, UK: Cambridge University Press; 2004

[45] Carcione JM. Wave propagation in anisotropic linear viscoelastic media: Theory and simulated wavefields. Geophysical Journal International. 1990; **101**(3):739-750

[46] Chadwick P. Wave propagation in transversely isotropic elastic media—I. Homogeneous plane waves. Proceedings of the Royal Society of London A. 1989; **422**(1862):23-66

[47] Chang SJ. Diffraction of plane dilatational waves by a finite crack. The Quarterly Journal of Mechanics and Applied Mathematics. 1971;**24**(4): 423-443

[48] Chattopadhyay A, Choudhury S. Magnetoelastic shear waves in an infinite self-reinforced plate. International Journal for Numerical and Analytical Methods in Geomechanics. 1995;**19**(4):289-304

[49] Das SC, Dey S. Edge waves under initial stress. Applied Scientific Research. 1970;**22**(1):382-389

[50] Fu YB. Existence and uniqueness of edge waves in a generally anisotropic elastic plate. Quarterly Journal of Mechanics and Applied Mathematics. 2003;**56**(4):605-616

[51] Fu YB, Brookes DW. Edge waves in asymmetrically laminated plates. Journal of the Mechanics and Physics of Solids. 2006;**54**(1):1-21

[52] Sneddon IN. The distribution of stress in the neighbourhood of a crack in an elastic solid. Proceedings of the Royal Society A. 1946;**187**(1009):229-260

[53] Sengupta PR, Nath S. Surface waves in fibre-reinforced anisotropic elastic media. Sadhana. 2001;**26**(4):363-370

[54] Shearer PM. Introduction to Seismology. 2nd ed. Cambridge: Cambridge University Press; 2009

[55] Sheriff RE, Geldart LP. Exploration Seismology. Cambridge, New York: Cambridge University Press; 1995

[56] Kolsky H. Stress Waves in Solids. New York: Dover Publication; 1963

[57] Gubbins D. Seismology and Plate Tectonics. Cambridge, New York: Cambridge University Press; 1990

[58] Kaplunov J, Prikazchikov DA, Rogerson GA. On three-dimensional edge waves in semi-infinite isotropic plates subject to mixed face boundary conditions. The Journal of the Acoustical Society of America. 2005;**118**(5): 2975-2983

[59] Kaplunov J, Pichugin AV, Zernov V. Extensional edge modes in elastic plates and shells. The Journal of the Acoustical Society of America. 2009;**125**(2): 621-623

[60] Pichugin AV, Rogerson GA. Extensional edge waves in pre-stressed incompressible plates. Mathematics and Mechanics of Solids. 2012;**17**(1):27-42

[61] Hool GA, Kinne WS, Zipprodt RR. Reinforced Concrete and Masonry Structures. New York: McGraw-Hill; 1924

[62] Spencer AJM. Boundary layers in highly anisotropic plane elasticity. International Journal of Solids and Structures. 1974;**10**(10):1103-1123

Section 2

Engineering Applications in Theory of Elasticity

Chapter 3

FEA and Experimentally Determination of Applied Elasticity Problem for Fabricating Aspheric Surfaces

Abstract

The elastic deformation machining method is suitable for fabricating aspheric surfaces that have excellent physical properties of elastic materials. The machining process is carried out with the deformation model without mold. When vacuum pressure is supplied to the workpiece, the top surfaces of workpiece are deformed into aspheric shape. After machining process, the bottom surface will be formed into the aspheric shape and the top surface returns to its original flat surface form due to internal force and bending moments of the material. However, the accuracy will decrease due to the reduced thickness while the vacuum pressure keeps unchanged during machining process. Therefore, it is necessary to carry out the finite element analysis (FEA) to determine the vacuum pressure with corresponding to the reduced thickness. In addition, the mold with its surface approximates and the desired surface form of the lens is also presented. When uniform vacuum pressure is supplied to the workpiece through small holes of the mold, the workpiece will be deformed into aspheric profile as similar to the mold surfaces. In order to improving the form accuracy, the FEA and the experiment are studied for modifying the mold profile to correspond with bending strength of workpiece material.

Keywords: elastic deformation, aspheric surface, glass lapping, glass molding, vacuum pressure, experimental study, finite element analysis

1. Introduction

Nowadays, for managing laser light in sophisticated and compact laser systems, aspheric lenses are the most powerful lenses. In these systems, it is generally accepted that spherical aberration is the most common performance detractor. From the use of spherical surfaces, it is found that they artificially limit focusing and collimating accuracy. In spite of the fact that spherical geometry is not optimal for refracting light that has been known for centuries, the high cost and difficulty of fabricating nonspherical (aspheric) surfaces has inhibited them from a wider use.

Because aspheric surfaces offer advantages such as high resolution, light weight, and low cost, they are widely used in the opto-electronics industry. As aspheric surfaces are more effective in shaping the light than spherical surfaces, they have recently been used in measurement instruments, astronomy, and optical lens [1, 2]. **Figure 1** shows some applications which employ aspheric surfaces.

In most general terms, an optical lens can be determined as a refracting device that reconfigures the light wave front incident upon it. The phase, direction of propagation, intensity, and polarization state are the properties of the incident light beam which are influenced by a lens. Surface form and roughness, diameter, subsurface defects generated during the fabrication process, shape accuracy, physical and mechanical properties of the optical material, and other optical conditions, such as the angle of incidence of light beam, absorption, reflection, and environmental influences, are some of the major characteristics that govern the performance of an optical lens [3].

To overcome the aberration problems of spherical lenses, a number of spherical surfaces with different signs of aberrations have to be utilized to balance and minimize the final aberration to obtain high quality images. In principle, the optical system designer can always use enough spherical lenses to simultaneously correct for all of the common optical aberrations in a lens system if the number of elements used in an optical system is not limited. The number of surfaces required to do this may be so large that the resulting lens assembly is excessively large in size and weight, and expensive to produce. In addition, the transmission of the assembly lens may be unduly reduced due to the residual reflections from each surface, and the bulk absorption in each lens.

The usage of aspheric surfaces, both with and without the incorporation of diffractive elements, allows the design and construction of assembly lens with the same or even better optical performance than an equivalent all-spherical system. However, in most cases, with a significant reduction in the number of elements required, there is a significant improvement in the overall lens assembly size, weight, cost, and optical transmission. In many cases, in an optical system, each aspheric surface can be applied to replace at least two other spherical surfaces. Hence, aspheric lenses are more efficient because additional error-correcting lenses are not required. **Figure 2** is an illustration of spherical and aspheric lens systems.

Cutting techniques such as turning and milling processes are usually utilized for the production of aspheric glass lenses as shown in **Figure 3**.

The machining processes, which usually consist of computer numerically controlled (CNC) generators, are employed to machine an aspheric shape on a lens to generate the desired shape. In glass machining, the roughness on a cutting edge has a larger effect on surface finish than that of metal machining. Glass workpiece can be machined without brittle fractures with an undeformed chip thickness less than 1 μm in milling and turning processes [4–7].

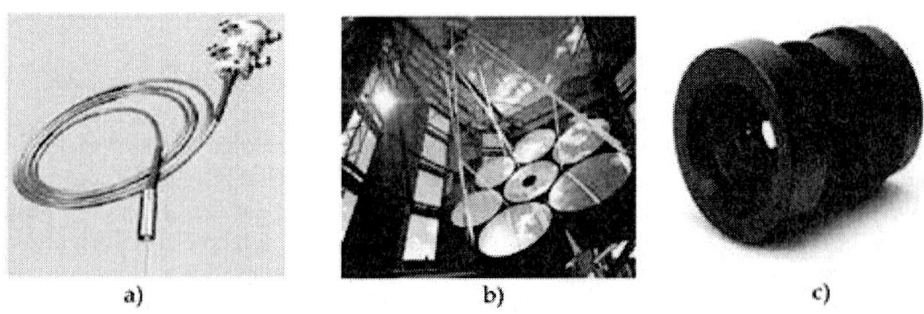

Figure 1.
Application of aspheric surfaces: (a) measurement instruments, (b) astronomy, and (c) optical lens.

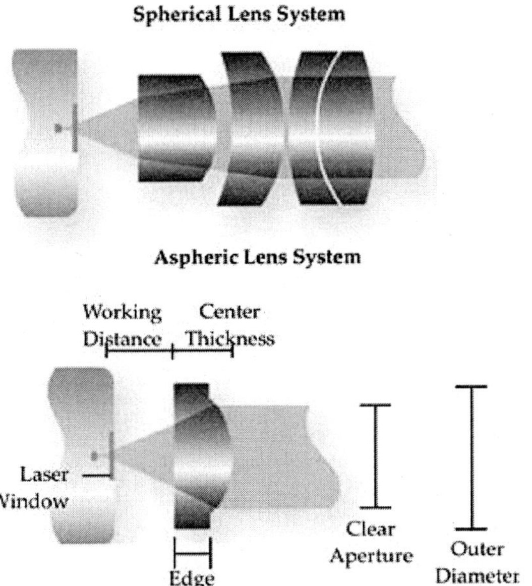

Figure 2.
Spherical vs. aspheric lens systems.

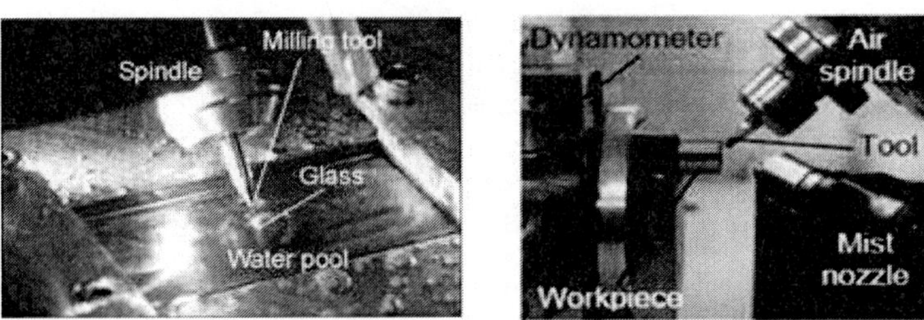

Figure 3.
Schematic illustration of milling and turning processes.

Thereafter, the optical lenses are fine machined by grinding, and then followed by polishing to achieve the good surfaces. In the grinding process, if the depth of cut is below a certain value, the material removal mode is ductile flow which is characterized by low surface roughness and subsurface damage [8–12]. **Figure 4** is an illustration of a precision grinding process.

"Precessions" polishing is an automated polishing method that uses a 7-axis CNC machine tool for polishing spherical and aspheric surfaces [13, 14]. Based on contact between the workpiece surface and polishing tool, the polishing spot of desired size is generated by controlling the load cell in polishing process. The polishing tool then moves in angular steps around the local normal to the part surface during machining process. The 7-axis CNC capability of the machine also makes the generation of free-form surfaces possible. **Figure 5** shows a schematic illustration of a "Precessions" polishing process.

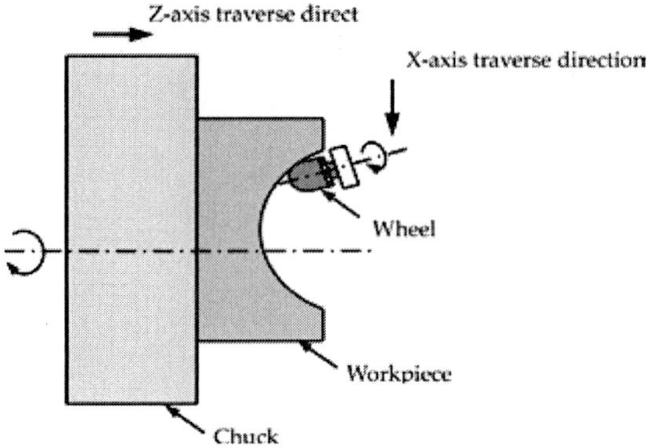

Figure 4.
Schematic illustration of a precision grinding process.

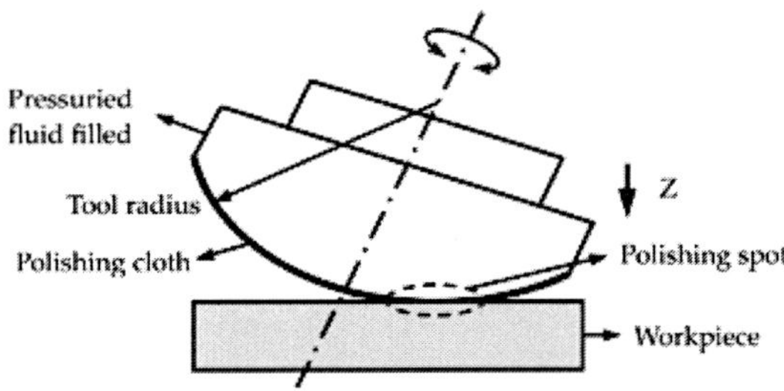

Figure 5.
Schematic illustration of a "Precessions" polishing process.

To fabricate aspheric surfaces, the movement of the tool must be constrained in the machining process. A sub-aperture tool (smaller in size than the lens) on a modified polishing machine is then utilized, and by controlling the amount of time the tool spends working at a given lens location, a desired aspheric surface can be fabricated. In addition to the complexity of the machining processes, conventional aspheric fabrication is highly sensitive to the manufacturing conditions, which strongly depend on the positioning accuracy of the machine, the condition of the grinding wheel, and the vibrations in the system. These factors result in an expensive manufacturing cost and a low production yield.

Compared to traditional cold-working methods, glass molding and precision injection molding have greatly advanced the fabrication technologies for aspheric lens industry because of their unique advantages such as excellent compatibility, high efficiency, great flexibility, and high consistency [15–17]. The mass production of aspheric glass lenses is fabricated by applying the technologies. In the glass molding technique, a glass lens is fabricated by compressing glass melting at a high temperature and replicating the shapes of the mold without any need of further machining. **Figure 6a** shows the process begins by putting a glass gob on top of a lower mold. Both the glass gob and the mold are heated to a molding temperature above the transition temperature of glass (**Figure 6b**). After the glass and the mold temperature have reached a steady state molding temperature, the mold is closed by moving the lower mold (**Figure 6c**). The temperature is maintained during the molding step. All steps are performed in vacuum environment. Then, by holding

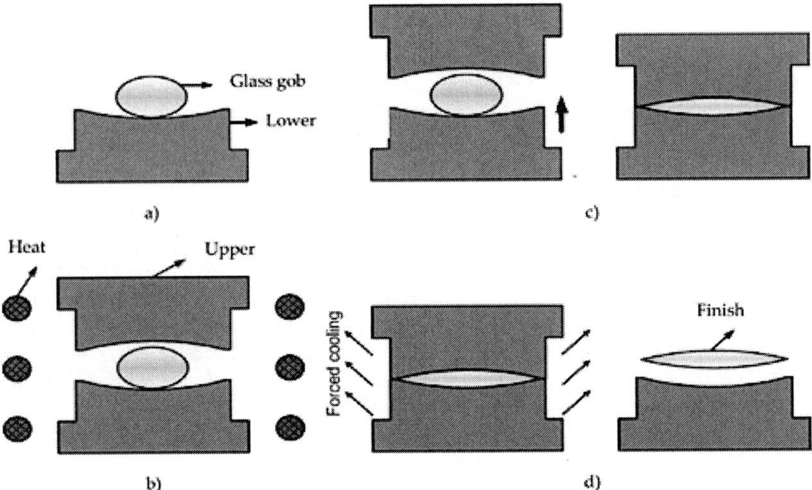

Figure 6.
Schematic illustration of a lens molding process: (a) Molds and glass gob, (b) Heating, (c) Heating and pressing, and (d) Cooling and release.

the pressed load for a short time at a slow cooling rate, the stress in the glass lens is relaxed. Lastly, the formed glass lens is rapidly cooled to ambient temperature and released from the molds (**Figure 6d**). A BK7 glass fabricated using this molding process has surface roughness of approximately 5 nm *Ra*, and form accuracy of 0.2 μm *P-V*.

In spite of the obvious advantages, there are serious drawbacks that currently limit the application of injection molding and glass molding technologies to smaller size aspheric lens fabrications. A typical drawback is the altering of optical properties such as refractive index, due to heating and annealing of the glass material, and the uneven shrinking due to the cooling process that causes error in lens profile [18].

In contrast, the elastic deformation machining method is a good technique that the workpiece will be deformed into aspheric shape prior to the lapping process under vacuum pressure. While the vacuum pressure is remained, the opposite side is polished to optical flatness by the lapping wheel. When the vacuum pressure is released, the bottom surface of the workpiece will be shaped into an aspheric shape and the top surface will restore to its flat surface form by internal force and bending moments. Consequently, for machining materials which have excellent physical properties due to their perfect crystal structures, the elastic deformation method is appropriate [19].

2. Theory of elastic deformation

Based on the elasticity of the material, the circular flat plate is deformed to an aspheric surface by applying the pressure in the elastic deformation machining method. The deflection of the circular plate can be calculated by using appropriate plate theory. There are two types of edge support for circular plate, such as fixed (or clamped) edge and simply supported edge which are considered in this section.

2.1 Basic equations for circular plate in elastic deformation

The amount of deflection of circular plate can be determined by solving the differential equations of an appropriate plate theory [20]. Two types of edge support include clamped and simply supported edge which are used in the elastic deformation method. In the case of simple bending of circular plate, the amount of deflection w is assumed to be very small in comparison with plate thickness. According to the small deflection theory of thin homogenous elastic plates, the deformation in the middle plane of the plate can be neglected and the straight line initially normal to the middle surface to the plate remains straight. In addition, the stress (i.e., transverse normal stress) is small when compared to other stress components and should be neglected in stress-strain relationship. Under these conditions, the three dimensional plate problem can be reduced to two dimensions. The linear theory of elasticity can be used to derive the governing differential equation for a plate subject to uniform transverse loads. The equation for small deformation w of a thin circular plate of constant thickness h is:

$$D.\nabla^2(\nabla^2 \mathrm{w}) - p = 0; D = Eh^3/12(1-\nu^2) \tag{1}$$

E and v are the Young's modulus and Poisson's coefficient. D is the rigidity constant of the plate, and p is the load on the plate. Because of the rotational symmetry, the Laplacian operator ∇^2 in polar coordinates r and θ can be written as:

$$\nabla^4 w = \left(\frac{\partial^2}{\partial r^2} + \frac{1}{r}\frac{\partial}{\partial r} + \frac{1}{r^2}\frac{\partial^2}{\partial \theta^2}\right)\left(\frac{\partial^2 w}{\partial r^2} + \frac{1}{r}\frac{\partial w}{\partial r} + \frac{1}{r^2}\frac{\partial^2 w}{\partial \theta^2}\right) \tag{2}$$

The moment can be written in the form,

$$M_r = -D\left[\frac{\partial^2 w}{\partial r^2} + \nu\left(\frac{1}{r^2}\frac{\partial^2 w}{\partial \theta^2} + \frac{1}{r}\frac{\partial w}{\partial r}\right)\right] \tag{3}$$

$$M_t = -D\left[\frac{1}{r}\frac{\partial w}{\partial r} + \frac{1}{r^2}\frac{\partial^2 w}{\partial \theta^2} + \nu\frac{\partial^2 w}{\partial r^2}\right] \tag{4}$$

where M_r and M_t are radial moment and tangential moment.

If the load acting on the plate is symmetrically distributed about the axis perpendicular to the middle plane of the plate, the deflection w is independent of θ, when Eqs. (3) and (4) becomes:

$$\left(\frac{d}{dr^2} + \frac{1}{r}\frac{d}{dr}\right)\left(\frac{d^2 w}{dr^2} + \frac{1}{r}\frac{dw}{dr}\right) = \frac{p}{\mathrm{D}} \tag{5}$$

In other form, it shown as

$$\left(\frac{d^2 w}{dr^2} + \frac{1}{r}\frac{dw}{dr}\right) = \frac{1}{r}\frac{d}{dr}\left(r\frac{dw}{dr}\right) \tag{6}$$

Eq. (5) can be written as

$$\frac{1}{r}\frac{d}{dr}\left\{r\frac{d}{dr}\left[\frac{1}{r}\frac{d}{dr}\left(r\frac{dw}{dr}\right)\right]\right\} = \frac{p}{D} \tag{7}$$

Multiply both sides of Eq. (6) by r and then integrate to obtain,

$$r\frac{d}{dr}\left[\frac{1}{r}\frac{d}{dr}\left(r\frac{dw}{dr}\right)\right] = \frac{pr^2}{2D} + C_1 \tag{8}$$

or,

$$\frac{d}{dr}\left[\frac{1}{r}\frac{d}{dr}\left(r\frac{dw}{dr}\right)\right] = \frac{pr}{2D} + \frac{C_1}{r} \tag{9}$$

By successive integrations, the deflection can arrive finally at

$$w_r = \frac{pr^4}{64D} + C_1 r^2 \ln r + C_2 r^2 + C_3 \ln r + C_4 \tag{10}$$

The shears for the symmetrically loaded plate can be given as follow,

$$\frac{d}{dr}\left(\frac{d^2w}{dr^2} + \frac{1}{r}\frac{dw}{dr}\right) = -\frac{Q}{D} \tag{11}$$

and from Eq. (9),

$$\frac{pr}{2D} + \frac{C_1}{r} = -\frac{Q}{D} \tag{12}$$

This equation indicates that Q would approach infinity as r approaches zero. To prevent this from happening, we make $C_1 = 0$. From Eq. (10), it can be seen that w becomes infinity at $r = 0$. To avoid this, the constant C_3 must be zero. Thus,

$$w_r = \frac{pr^4}{64D} + C_2 r^2 + C_4 \tag{13}$$

From Eq. (13), the amount of deflection w_r is the function of r in cylindrical coordinate system.

2.2 Circular plate with simply supported edge

The boundary conditions are $w = 0$ and $M_r = 0$ at $r = a$.
Eq. (13) can be written,

$$\frac{pa^4}{64D} + C_2 a^2 + C_4 = 0 \tag{14}$$

and

$$\frac{pa^2}{64}(12 + 4\nu) + \frac{C_2}{2}(1 + \nu) = 0 \tag{15}$$

From the equations, we can find

$$C_2 = -\frac{pa^2}{32D}\frac{3+\nu}{1+\nu};\ C_4 = \frac{pa^4}{64D}\frac{5+\nu}{1+\nu} \tag{16}$$

so that the deflection of every radial location can be calculated using,

$$w_r = \frac{p(a^2 - r^2)}{64D}\left[\frac{5+\nu}{1+\nu}a^2 - r^2\right] \tag{17}$$

The maximum deflection which occurs at r = 0, is given by,

$$w_0 = \frac{pa^4}{64D}\left(\frac{5+\nu}{1+\nu}\right) \tag{18}$$

Substitute $D = Eh^3/12(1 - v^2)$ into Eq. (18), we have:

$$w_0 = \frac{3p(1-\nu)(5+\nu)a}{16E}\left(\frac{a}{h}\right)^3 \tag{19}$$

From Eq. (19), we can see that the maximum deflection of the circular plate is relative of the diameter a and the ratio between diameter a and thickness h.

3. Elastic deformation machining method without mold

Figure 7 shows a schematic illustration of a lens elastic deformation process without mold.

Two surfaces of the workpiece are polished to certain flatness before fabricating as shown in **Figure 7a**. When vacuum pressure is supplied to the workpiece through

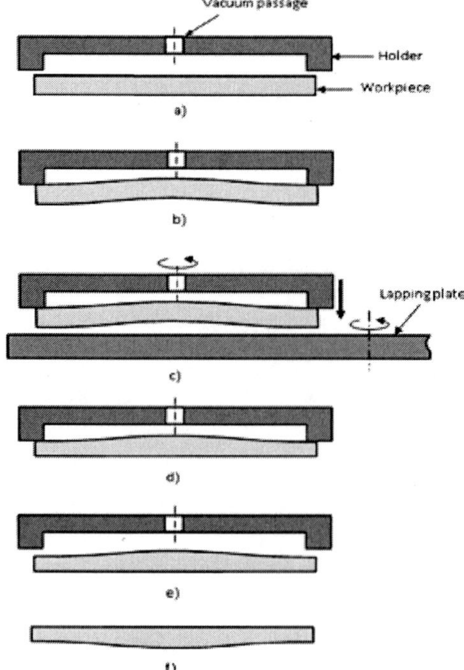

Figure 7.
(a–f) Schematic illustration of an aspheric surface elastic deformation process.

a hole, the workpiece is deformed in the middle. The edge of the workpiece is supported by the holder; therefore, it will be not moved. This makes the workpiece become a formed aspheric shape as presented in **Figure 7b**. The deflection of the workpiece can be calculated by using theoretical equation in boundary conditions of circular plate with simply supported edge. While the vacuum pressure is still remained, the workpiece and the holder start rotating and moving downward in contact with the lapping plate. Its opposite side will be polished to optical flatness as illustrated in **Figure 7c** and **d**. Then, the vacuum pressure is not supplied and the workpiece is also released from the holder as shown in **Figure 7e**. According to the **Figure 7f**, the bottom surface will be formed into the aspheric shape and the top surface returns to its original flat surface form due to material elasticity. It can be seen that the deformed workpiece surfaces can be restored by internal force and bending moments which are created from the vacuum pressure during machining process.

3.1 Finite element analysis for elastic deformation machining process

The vacuum pressure affects an amount of elastic deformation of the workpiece; hence, the accuracy of manufactured profile will also be highly dependent on the vacuum pressure as well. **Figure 8a** and **b** illustrates that the workpiece is lapped and polished to a flat surface while the vacuum pressure stays it at the initial deformed state.

The manufactured workpiece accuracy can be improved by adjusting the vacuum pressure during the machining process because of the changed workpiece thickness [21]. The vacuum pressure is defined by finite element analysis (FEA) results because theoretical calculation for complex surface is more difficult. In simulation process, a circular plate B270 optical glass with the edge supported by the holding device is listed in **Table 1**.

All elements of modeling were created by meshing with A20-node quadratic brick elements in reduced integration (C3D20R). **Figure 9** demonstrates the finite element model as follows.

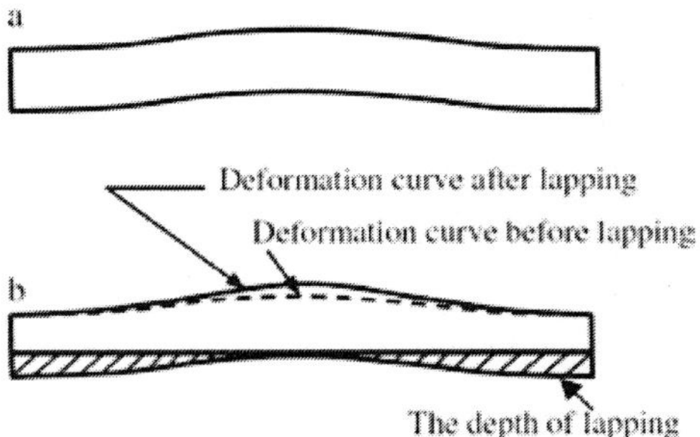

Figure 8.
(a and b) The deforming and lapping processes of glass plate. (a) The glass plate is deformed before lapping and (b) the glass plate is deformed in lapping.

Density (kg/m^3)	Young's modulus (GPa)	Knoop hardness HK100 (kg/mm^2)	Poisson ratio
2550	71.5	542	0.22

Table 1.
Material properties of B270 optical glass [22].

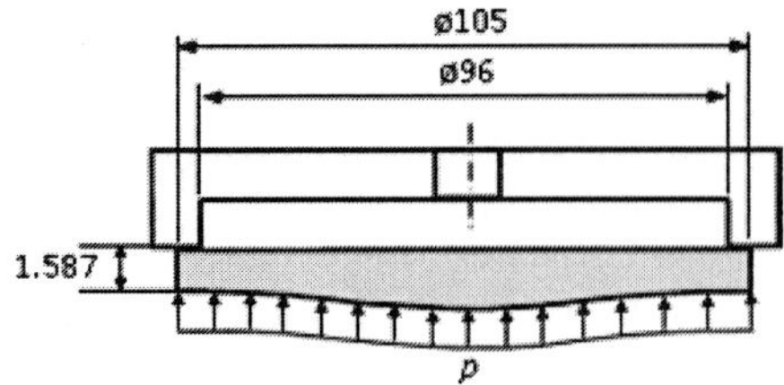

Figure 9.
Simulation model of workpiece.

When the vacuum pressure keeps unchanged, the workpiece thickness is reduced during the lapping process. Therefore, the surface form of a glass plate will have some errors compared to desired surface form at the end of the machining process. The results of the FEA indicate that the deflection of workpiece is greater than desired curve. In order to enhance its accuracy, the vacuum pressure should be fixed at 42 kPa as shown in **Figure 10**.

3.2 Experimental setup

The B270 optical glass which is a clear, high transmission and high purity raw materials is chosen in this experiment. The workpiece sides are lapped and polished to flat surfaces. The lapping process is through the relative motion between the lapping plate and the workpiece, affected by abrasive slurry under distribution load. The silicon carbide (SiC) and cerium oxide (CeO_2) abrasive grain slurry are used in the experiment. The principle of lapping process can be seen in **Figure 11**.

In lapping process, a rigid iron surface covered by a flannelette plate is moved under the load on the glass surface, with abrasive particles suspended in water between them. **Table 2** demonstrates parameters for the machining process. To remove microcrack layer and trace after the lapping process, a polishing step is required. This step is also carried out with the Nanopoli-100 precision polishing machine. The polishing parameters fixed unchanged as that in the initial lapping step, except that the abrasive is changed from SiC to CeO_2 as a fine polishing step.

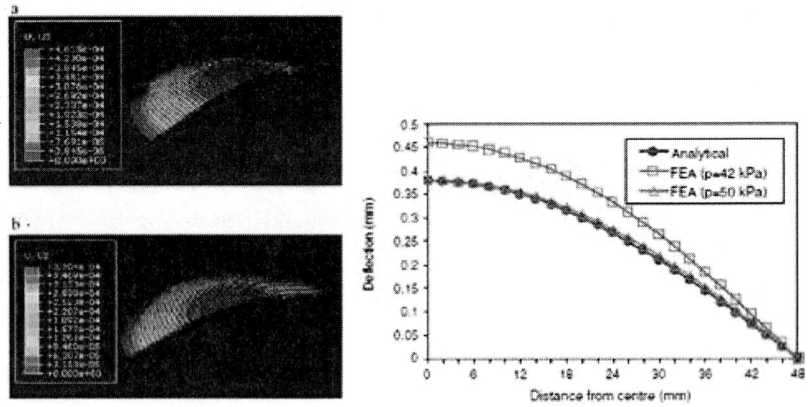

Figure 10.
Finite element analysis results and analytical results: (a) P = 50 kPa and (b) P = 42 kPa.

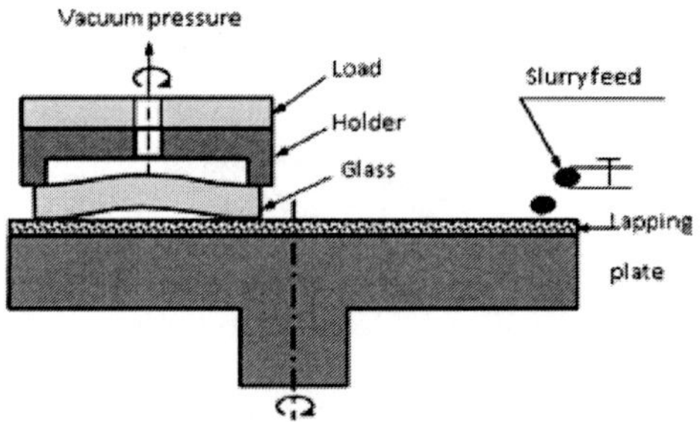

Figure 11.
Principle of lapping and polishing process.

Items	Lapping	Polishing
Abrasive	#1000 SiC	#10,000 CeO_2
Abrasive concentration in slurry (wt%)	10%	10%
Machining load (N)	20	15
Rotating speed of lapping plate (rpm)	60	40
Machining time (min)	240	60

Table 2.
Lapping and polishing conditions.

3.3 Experimental results

The component accuracy can be improved by adjusting the vacuum pressure values to compensate for its lost thickness during the lapping step. The vacuum pressure is defined through FEA results. **Figure 10** shows that the deformation curve of the workpiece is close to the desired curve when the vacuum pressure is fixed at 42 kPa. Therefore, the vacuum pressure should be reduced from 50 to 42 kPa in the experiment. **Figure 12** illustrates the deflection and deviation results of the experimental results and the theoretical calculations.

Depending on reducing pressure from 50 to 42 kPa and keeping stable through the entire lapping step, the experimental results agree greatly with theoretical calculations. The peak-valley value is reached at 1.6 μm.

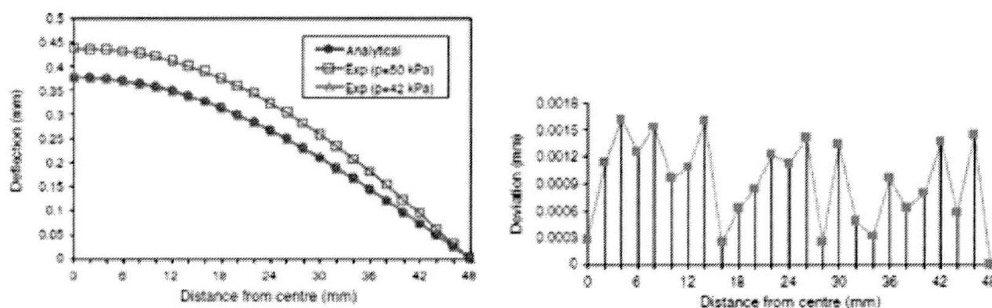

Figure 12.
The experimental results of deflection and deviation against theoretical results.

4. Elastic deformation machining method with mold

In the elastic deformation machining process without mold, the thickness of the plate is reduced while the vacuum pressure remains unchanged. Thus, the work-piece deformation to increase as lapping progresses. This will cause large deviation in surface form between finished workpiece and theoretical calculation. The mold with its surface approximates the desired surface form of the lens which is used for improving the machining precision. When vacuum pressure is supplied, the top surface of the workpiece will be deformed and then contacts the molded surface. **Figure 13** shows the basic concept of elastic deformation molding process [23].

The mold and workpiece surfaces are polished to flatness before fabricating as shown in **Figure 13a**. When uniform vacuum pressure is supplied to the workpiece through small holes of the mold, the workpiece will be deformed and then contacted with the aspheric surface of the mold as presented in **Figure 13b**. While deformed workpiece is kept stable under vacuum pressure, the bottom side of the workpiece is polished to flatness as illustrated in **Figure 13c** and **d**. Then, the vacuum pressure is not supplied; hence, the lapped side of workpiece will be formed into the mold surface while the opposite surface returns to its original flatness surface due to material elasticity as shown in **Figure 13e**.

4.1 Finite element analysis for elastic deformation machining process with mold

The standard aspheric formula is:

$$Z = \frac{cr^2}{1+\sqrt{1-(1+k)c^2r^2}} + \sum_{i=2}^{n} A_{2i}r^{2i} \tag{20}$$

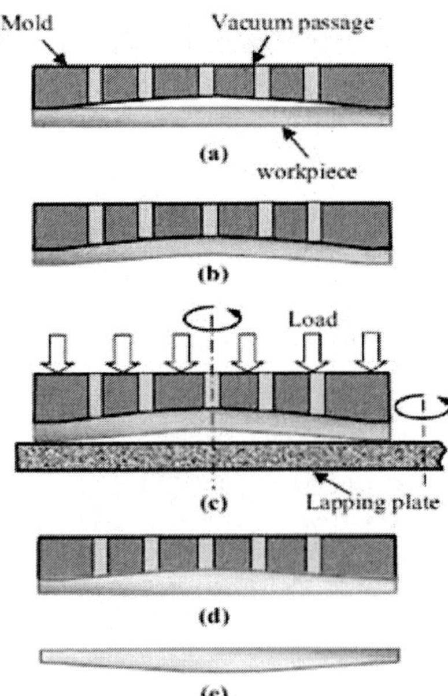

Figure 13.
Basic principle of elastic deformation molding process.

where Z, depth or "Sag" of the curve; r, distance from the center; c, curvature (=1/radius); K, conic constant; and A_{2i}, higher order terms.

The radius (R) is used for determining the aspheric terms such as their shallow or depth. The closest spherical surface is the radius which reaches the aspheric sag at the largest useful diameter [24]. **Figure 14** illustrates the aspheric lens sag.

An aspherical surface is built by using spherical surface combined with the higher order terms. Most optical designers use only the even-order terms from A2 to A20. The conic constant K has been used to design the initial aspheric, simple paraboloid and hyperboloid (as shown in **Table 3**).

In elastic deformation machining method, the accuracy of aspheric lens depends on the ability of elastic deformation and completely contacting the mold surfaces. The mold surface is defined by choosing the closest spherical surface (as shown in **Figure 15**). The FEA is designed for establishing the spherical surface through a simulation of contacting process between workpiece and mold surface.

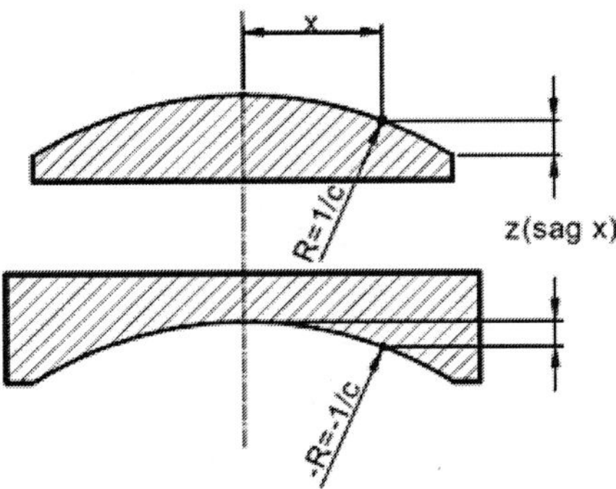

Figure 14.
The sag of aspheric lens.

Conic constant	Surface type
K = 0	Spherical
K = −1	Paraboloid
K < −1	Hyperboloid
−1 < K < 0	Ellipsoid
K > 0	Oblate ellipsoid

Table 3.
The relationship between conic constants and surface types.

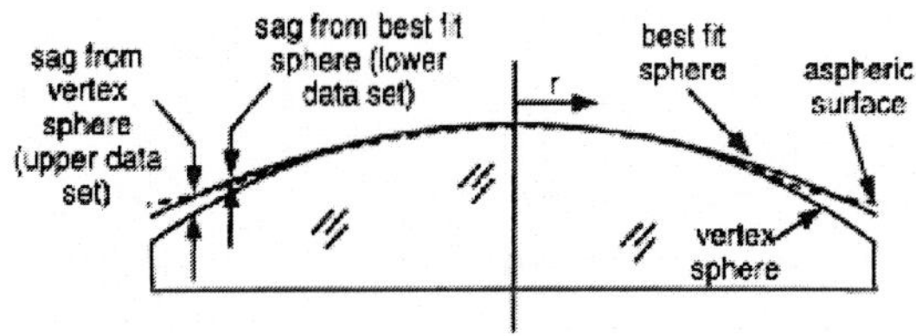

Figure 15.
The aspheric surface from best fit sphere.

In the simulation process, the thickness (h = 1.0 mm) and diameter (D = 50 mm) of the workpiece are suggested. In addition, the radius (R = 2500 mm) of spherical surface is chosen. The parameters of FEA model can be seen in **Figure 16a.**

The axisymmetric model is selected in this simulation process. The mold is chosen as an analytical rigid shell and the workpiece is a deformable shell. The analytical step of model is "Dynamic, Explicit". The interaction and the contact property are "Surface to surface contact" and "Penalty contact method," respectively. All elements of the workpiece are divided in meshing with A4-node bilinear axisymmetric quadrilateral elements in reduced integration. The workpiece mesh and boundary conditions are described in **Figure 16b**. The values of uniform vacuum pressure are opted in range of −80 to −100 kPa. The conic constant K = 0.25 is selected for the simulation process.

It is clear to see that **Figure 17** shows the deflection and deviation of workpiece under different vacuum pressures with the conic constant K = −3.

According to the results, the model with the conic constant K = −3, gives the best one and the deviation between the workpiece and the mold is the smallest. The workpiece and the mold can reach the best when the vacuum pressure approximates −95 kPa. However, when the vacuum pressure is larger than −95 kPa, the deviation results are still stable. Therefore, the conic constant K = −3 recommends for defining the aspheric surface of the mold.

The accuracy one can be innovated by modifying the mold profile to adopt with bending stress of workpiece material. This mold profile is redesigned by using the profile of workpiece after the deformed stage. An axisymmetric FEM model is established, and it consists of the new mold and workpiece. The uniform vacuum pressure is chosen as −95 kPa. The mold surface is redesigned with the conic constant K = −3 as displayed in **Figure 18.**

It can be noted that **Figure 19a** and **b** presents the deflection and deviation results between the workpiece and the new mold under supplied vacuum pressures, P = −95 kPa and K = −3.

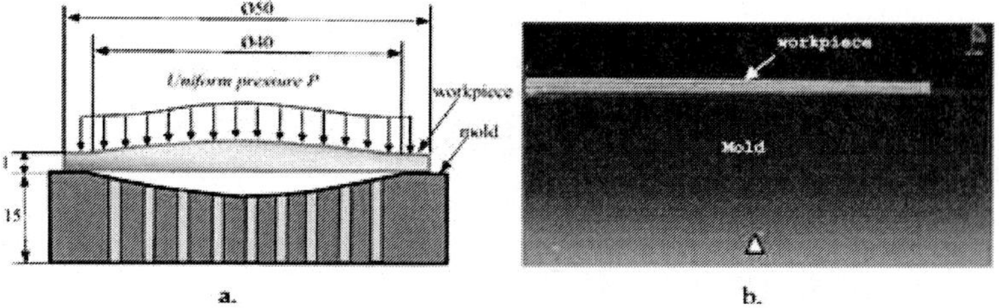

Figure 16.
(a) The simulation model and (b) FEM simulation model.

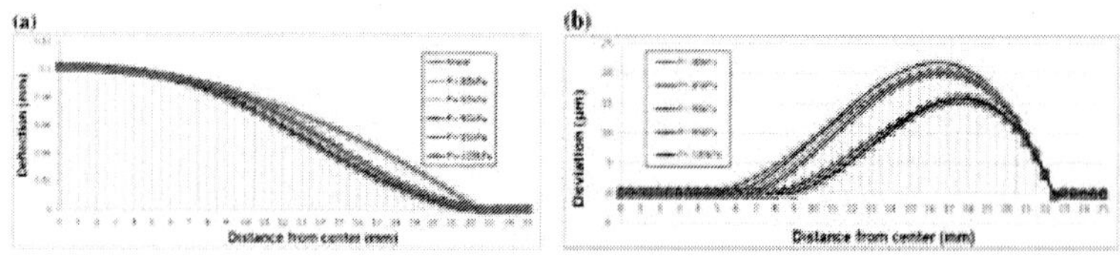

Figure 17.
(a and b) Deflection and deviation under different vacuum pressures (K = −3).

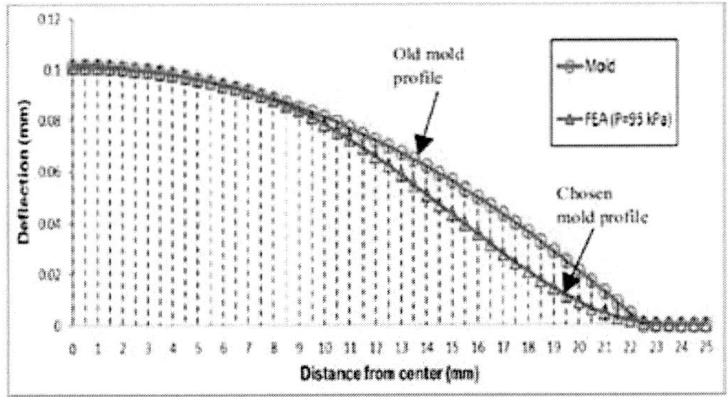

Figure 18.
The modified mold is chosen.

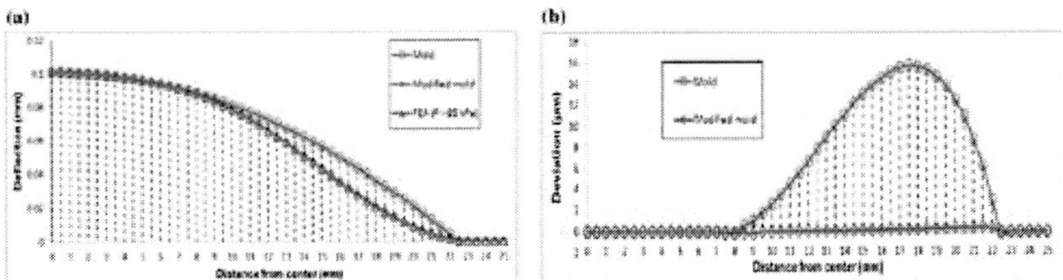

Figure 19.
(a and b) Deflection and deviation results between the workpiece and modified mold.

The form accuracy of workpiece is enhanced by using the new mold surface. The maximum deviation is less than *P-V* 0.35 μm while the former mold is about 15.02 μm.

4.2 Experimental setup

Figure 20 presents that the experiment was conducted to a precision polishing machine Preci-Polish 300. The B270 glass with a diameter of 50 mm and a thickness

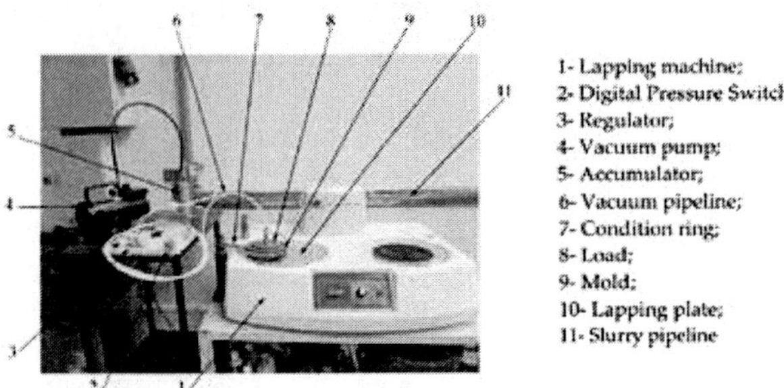

Figure 20.
Experimental set-up in the lapping processes. 1-Lapping machine; 2-digital pressure switch; 3-regulator; 4-vacuum pump. 5-accumulator; 6-vacuum pipeline; 7-condition ring; 8-load; 9-mold; 10-lapping plate; and 11-slurry pipeline.

of 1.0 mm is utilized in the experiment process. In addition, **Table 4** points the parameters for the machining process, in which the vacuum pressure is fixed as −95 kPa.

According to the simulation results, the new mold surface with the conic constant $K = -3$ is chosen. The multi-small holes of the modified mold surface are fabricated to fix the workpiece during the machining process. The new mold is machined on a precision CNC machining center.

4.3 Experimental results

Figure 21a and **b** shows that experimental results are compared to FEA with new mold surface under applied vacuum pressure $P = -95$ kPa.

Based on the experimental and FEA results, the deviation of workpiece is less than *P-V* 0.01 μm within the radius of about 12 mm. The maximum deviation is *P-V* 0.6 μm; however, the former mold is about 18.93 μm. It is clear to see that the experimental results agree greatly with FEA results. Therefore, the form accuracy of the workpiece is significantly improved when the new mold profile is redesigned according to the FEA results with $P = -95$ kPa and $K = -3$.

Items	Lapping	Polishing
Abrasive	#1000 SiC	#10,000 CeO_2
Abrasive concentration in slurry (wt%)	10%	10%
Machining load (N)	30	20
Rotating speed of lapping plate (rpm)	60	40
Machining time (min)	120	30

Table 4.
Lapping and polishing parameters.

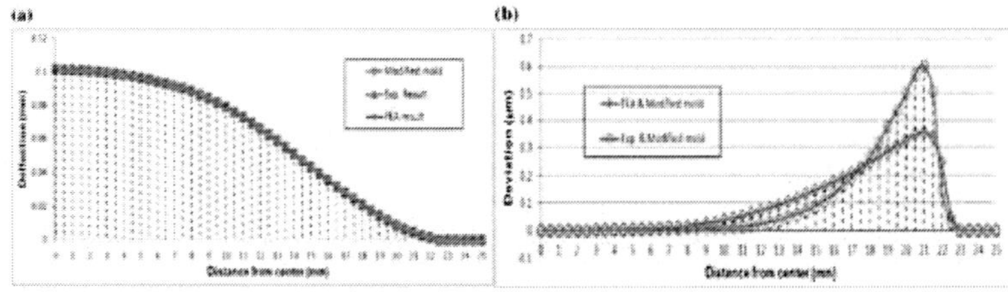

Figure 21.
(a and b) Experimental and FEA results with modified mold surface.

5. Conclusions

Based on the elasticity of the material, the elastic deformation machining is a method in which the vacuum pressure is used for fabricating complex aspheric surfaces. The amount of deflection of circular plate can be determined by solving the differential equations of an appropriate plate theory. The workpiece will be deformed into aspheric shape prior to the lapping process under the vacuum pressure. While the vacuum pressure is remained, the opposite side is polished to optical flatness by the lapping wheel. Then, the vacuum pressure is not supplied and hence, the bottom surface will be formed into the aspheric shape and the top surface will be restored to its flat surface form. Therefore, the method is suitable for manufacturing of optical lens with large aperture and low thickness glass materials.

In the elastic deformation machining process without mold, the manufactured workpiece accuracy can be increased by adjusting the vacuum pressure during the machining process because of the changed workpiece thickness. The vacuum pressure is defined through FEA results. According to the FEA, the deformation curve of the workpiece is reached to the desired curve when the vacuum pressure is fixed at 42 kPa. Depending on reducing the vacuum pressure from 50 to 42 kPa and keeping stable through the entire machining process, the experimental results agree greatly with theoretical calculations. The best peak-valley value *P-V* 1.6 µm was achieved in this method.

In order to achieve form accuracy of the workpiece in the elastic deformation machining process with mold, the mold with its surface approximates the desired surface form of the lens which is used for improving the machining precision. The accuracy one can be innovated by modifying the mold profile to adopt with bending stress of workpiece material. This mold profile is redesigned by using the profile of workpiece after the deformed stage. According to the simulation results, the new mold surface with the conic constant $K = -3$ and vacuum pressure $P = -95$ kPa are used for the experimental process. In this case, the deviation of workpiece is less than *P-V* 0.01 µm within the radius of about 12 mm. The maximum deviation is *P-V* 0.6 µm; however, the former mold is about 18.93 µm. It is clear to see that the experimental results agree greatly with FEA results.

References

[1] Aono Y, Negishi M, Takano J. Development of large-aperture aspherical lens with glass molding. Advanced Optical Manufacturing and Testing Technology. 2000;**4231**:16-23. DOI: 10.1117/12.402759

[2] Suzuki H, Highuchi O, Huriuchi H, Shibutani H. Precision cutting of micro-axis-symmetric spherical surface with 3-axes controlled diamond tool. In: Proceedings of the Second Euspen International Conference on European Society for Precision Engineering and Nanotechnology; 27-31 May 2001; Italy. pp. 844-847

[3] Anurag J. Experimental study and numerical analysis of compression molding process for manufacturing precision aspherical glass lenses [thesis]. Columbus: Ohio State University; 2006

[4] Takashi M, Takenori O. Cutting process of glass with inclined ball end mill. Journal of Materials Processing Technology. 2008;**200**:356-363. DOI: 10.1016/j.jmatprotec.2007.08.067

[5] Suzuki H, Moriwaki T, Yamamoto Y, Goto Y. Precision cutting of aspherical ceramic molds with micro PCD milling tool. Annals of the CIRP. 2007;**56**: 131-134. DOI: 10.1016/j.cirp. 2007.05.033

[6] Kim HS, Kim EJ, Song BS. Diamond turning of large off-axis aspheric mirrors using a fast tool servo with on machine measurement. Journal of Materials Processing Technology. 2004; **146**:349-355. DOI: 10.1016/j. jmatprotec.2003.11.028

[7] Li L, Yi AY, Huang C, Grewell DA, Benatar A, Chen Y. Fabrication of diffractive optics by use of slow tool servo diamond turning process. Optical Engineering. 2006;**45**:113401. DOI: 10.1117/1.2387142

[8] Bifano TG, Dow TA, Scattergood RO. Ductile-regime grinding: A new technology for machining brittle materials. Journal of Engineering for Industry. Transactions of the ASME. 1991;**113**:184-189. DOI: 10.1016/ 0141-6359(92)90161-O

[9] Lambropoulus JC, Fang T, Funkenbusch PD, Jacobs SD, Cumbo MJ, Golini D. Surface micro-roughness of optical glasses under deterministic micro-grinding. Applied Optics. 1996; **35**:4448-4462. DOI: 10.1364/ AO.35.004448

[10] Namba Y, Abe M, Kobayashi A. Ultra-precision grinding of optical glasses to produce super-smooth surfaces. Annals of the CIRP. 1993;**42**: 417-420. DOI: 10.1016/S0007-8506(07) 62475-5

[11] Chen WK, Kuriyagawa T, Huang H, Yoshihara N. Machining of micro aspherical mould inserts. Journal of Precision Engineering. 2005;**29**:315-323. DOI: 10.1016/j.precisioneng.2004.11.002

[12] Brinksmeier E, Mutlugunes Y, Klocke F. Ultra-precision grinding. CIRP Annals-Manufacturing Technology. 2010;**59**:652-671. DOI: 10.1016/j.cirp.2010.05.001

[13] Bingham R, Walker D, Kim D. Novel automated process for aspheric surfaces. In: Proceedings of SPIE – The International Society for Optical Engineering. 2000;**4093**:445-448. DOI: 10.1117/12.405237

[14] Walker D, Brooks D, King A. The "Precessions" tooling for polishing and figuring flat, spherical and aspheric surfaces. Optics Express. 2003;**11**: 958-964:958-964. DOI: 1364/ OE.11.000958

[15] Nelson J, Scordato M, Schwertz K. Precision lens molding of asphero

diffractive surfaces in chalcogenide materials. Optifab. 2015:96331L. DOI: 10.1117/12.2195764

[16] Mahajan P, Dora PT, Sandeep TS. Optimized design of optical surface of the mold in precision glass molding using the deviation approach. International Journal for Computational Methods in Engineering Science and Mechanics. 2015;**16**:53-64. DOI: 10.1080/15502287.2014.976677

[17] Yi AY, Tao B, Klocke F. Residual stresses in glass after molding and its influence on optical properties. Procedia Engineering. 2011;**19**:402-406. DOI: 10.1016/j.proeng.2011.11.132

[18] Kunz A. Aspheric freedoms of glass-precision glass moulding allows cost-effective fabrication of glass aspheres. Optik & Photonik. 2009;**4**:46-48. DOI: 10.1002/opph.201190063

[19] Mori Y, Yamamura K, Endo K, Yamauchi K, Yasutake K, Goto H, Kakiuchi H, Sano Y, Mimura H. Creation of perfect surfaces. Journal of Crystal Growth. 2005;**275**:39-50. DOI: 10.1016/j.jcrysgro.2004.10.097

[20] Ventsel E, Krauthammer T, editors. Thin Plates and Shells: Theory, Analysis, and Applications. 1st ed. Boca Raton: CRC Press; 2001. p. 688

[21] Nguyen DN, Lv BH, Yuan JL, Wu Z, Lu HZ. Experimental study on elastic deformation machining process for aspheric surface glass. International Journal of Advanced Manufacturing Technology. 2013;**65**:525-531. DOI: 10.1007/s00170-012-4191-3

[22] Schott. Schott B270 Super-white Properties [Internet]. 2015. Available from: https://psec.uchicago.edu/glass/SchottB270Properties-KnightOptical.pdf [Accessed: 24-05-2018]

[23] Nguyen DN. Study on improving the precision of form surface produced in elastic deformation molding process. International Journal of Advanced Manufacturing Technology. 2017;**93**:3473-3484. DOI: 10.1007/s00170-017-0766-3

[24] Kweon G, Kim CH. Aspherical lens design by using a numerical analysis. Journal of the Korean Physical Society. 2007;**51**:93-103. DOI: 10.3938/jkps.51.93

Chapter 4

Concept of Phase Transition Based on Elastic Systematics

Abstract

The use of elastic constants systematics to describe fundamental properties of engineering materials has made materials science education and its related subjects increasingly important not only for manufacturing engineers but also for mankind at large. In this chapter, we present actual scaling of phase transition-driven considerations, such as martensitic transformation and transformable shape memory formation via elastic constant systematics. The scaling in terms of the simple and polycrystals mechanical stability criteria based on the elastic moduli and an acoustic anisotropy is in good agreement with novel experimental data from the literatures, and further, a long-standing concern in predicting polycrystalline elastic constants was considered beyond the commonly encountered criteria.

Keywords: elastic, elastic modulus, martensitic transformation, shape memory effect, elastic constant, ductility criterion, mechanical properties

1. Introduction

The ingenuity and the art required to tailor precisely the desired physical and structural properties in materials have been the main goal of the material scientists and engineers. Elastic response (i.e. elastic constant) to an applied load is one of such basic properties of all solids and originates from the distortion of atomic bonds. Simply put, elastic constants are a reflection of the fundamental thermodynamic properties that take place in the crystal lattice of solids. Complementary to this, the otherwise inaccessible essential information can be revealed from their temperature and stress dependencies of these important constants. For instance, the crystal structures of the three long periods of transition elements change more or less systematically from hcp through bcc to fcc as their group numbers increase from IV to VIII as does their elastic properties. Thus, the knowledge of microscopic elasticity can provide a fruitful ground for the exploration of the material behaviour yet uncommon to our knowledge about the relationship between crystal structure and bonding.

The earliest foundation of elastic theory dates back to seventeenth century (around 1821), when Navier first gave the equation for the equilibrium and motion of elastic solids [1], but modern foundation of microscopic elastic theory was established by the work of Born and Huang [2], followed by other excellent treatments [3]. It is well known that crystalline solids are by no means ideal and invariably contain some lattice defects such as vacancies, solute atoms or some extent of disorder. These point defects strongly affect almost all properties of materials, including elastic behaviours. In effect, the early investigators of these phenomena were motivated by the response of naturally occurring anisotropic materials such as wood and other crystalline solids. On that premise, of interest here is the relationship between crystal structure and elastic properties, mainly because of the important information they provide about nature of binding forces in solids.

Over the past three decades, elastic constants of some simple crystals have been a subject of numerous researches and have been investigated both theoretically and experimentally. Some of the outcomes have revealed that fundamental elastic properties of a martensitic crystal are fully determined by the elastic constants C_{ij}. All macroscopic elastic moduli (Young's and shear modulus, Poisson ratio, etc.) can be derived from the C_{ij} at least within certain upper and lower bounds [4]. There is considerable evidence that the magnitude of $C^{'} = (C_{11} - C_{12})/2$ elastic shear modulus in metallic bcc structures is closely related to the occurrence of martensitic phase transformations and is thus a useful parameter for estimating bcc structures [5]. Similarly, whether a structural material shows plastic flow or brittle fracture on loading is of clear practical significance. Brittleness in polycrystalline metals can be intrinsic or induced. The basic question is: Do these two general properties (i.e. phase stability and elastic properties) of crystals correlate to each other?

2. Analytical criterion of elastic constants of perfect crystals

The elastic properties are among the most important physical properties of materials and the importance of studying elastic properties of materials cannot be overemphasised. The knowledge of elastic properties is essential for both structural design and experimental mechanics [6]. It also enables the assessment of the sufficiency of strength, stiffness and stability of newly developed materials. Although the crystals are assumed to free from lattice imperfections and difficult to produce, their study had always been the building block for a better understanding of the behaviour of bulk materials. Usually, the determination of elastic properties of crystalline solids is based on its single or perfect crystal configuration under special loading conditions. The elastic moduli are the material constants that connect stress with strain and are therefore crucial to engineering applications. A crystal subjected to external load undergoes dimensional change. If the eternal load is a stress tensor denoted by σ_{ij}, then the deformation per unit length in three-dimensional space, can be described by a strain tensor, e_{ij}. Within the elastic limit or for sufficiently small deformations, the stress tensor is a linear function of the strain tensor and the generalised delta notation of Hooke's law can be used to express the relationship between these two quantities [7] as:

$$\sigma_{ij} = C_{ijkl}e_{kl} \tag{1}$$

where C_{ijkl} is the proportionality constant that characterises the crystal's resistance to elastic shape change; often referred to as the elastic coefficients or elastic constants or elastic moduli or stress-strain coefficients [8].

The inverse relation between the strain and the stress can be determined by taking the inverse of stress-strain relation to get:

$$e_{ij} = S_{ijkl}\sigma_{kl} \tag{2}$$

Here, S_{ijkl} represents the elastic compliance of the crystal. From symmetry or equilibrium principles, the state of stress in an elastic body can be approximated by six independent stress and strain components. And as such the stress and strain components in Eq. (1) can be expressed in three orthogonal axes as:

$$
\begin{aligned}
\sigma_x &= C_{11}e_{xx} + C_{12}e_{yy} + C_{13}e_{zz} + C_{14}e_{yz} + C_{15}e_{zx} + C_{16}e_{xy} \\
\sigma_y &= C_{21}e_{xx} + C_{22}e_{yy} + C_{23}e_{zz} + C_{24}e_{yz} + C_{25}e_{zx} + C_{26}e_{xy} \\
\sigma_z &= C_{31}e_{xx} + C_{32}e_{yy} + C_{33}e_{zz} + C_{34}e_{yz} + C_{35}e_{zx} + C_{36}e_{xy} \\
\tau_{yz} &= C_{41}e_{xx} + C_{42}e_{yy} + C_{43}e_{zz} + C_{44}e_{yz} + C_{45}e_{zx} + C_{46}e_{xy} \\
\tau_{zx} &= C_{51}e_{xx} + C_{52}e_{yy} + C_{53}e_{zz} + C_{54}e_{yz} + C_{55}e_{zx} + C_{56}e_{xy} \\
\tau_{xy} &= C_{61}e_{xx} + C_{62}e_{yy} + C_{63}e_{zz} + C_{64}e_{yz} + C_{65}e_{zx} + C_{66}e_{xy}
\end{aligned} \tag{3}
$$

Here, e_{xx}, e_{yy} and e_{zz} are tensile strains, e_{xy}, e_{yz} and e_{zx} are shear strains. The experimental values of elastic constants, C_{ijkl}, were originally determined by considering the response of crystals to small strains or unstressed lattice using Eq. (1). Beyond using Eq. (1) based on measured stress-strain relations, there are now methods of determining elastic constants from the first principles often referred to as ab initio methods. There are many methods of evaluating elastic coefficients such as the one based on expanding the internal strain energy of the crystal [7]. Thus, we may write as Eq. (4),

$$U = U_0 + V_0\sum\sigma_i e_i + \frac{1}{2}V\sum_i\sum_i C_{ij}e_ie_j + ... \tag{4}$$

where U is the energy of the crystal, is a quadratic function of the strains, in the approximation of Hooke's law (recall the expression for the energy of a stretched spring). V_0 is its equilibrium volume and e denotes an elastic strain. If the material is a crystal, the number of independent elastic constants is reduced further depending on the crystal system.

Elastic coefficients and elastic moduli have significant effect of mechanical response of crystals. Elastic constants, $C_{ij}(C_{11}, C_{12}, C_{44})$ and elastic moduli such as bulk modulus (B), shear modulus (G), Young's modulus (E) influence mechanical response of crystals. For instance, the bulk modulus (B) is associated with the hardness of materials which is of extreme importance in high-temperature and pressure applications, while elastic constants could provide essential information about bonding between adjacent atomic planes, anisotropic character of bonding and structural stability [7]. By far, the most widely reported elastic properties are E, G and B, corresponding to tensile, shear and hydrostatic loading, respectively. Since B signifies the compressibility of a substance, it can be calculated from the partial derivative of volume (V) and pressure (P) at constant temperature (T), as per Eq. (5).

$$B = \left({}^{\delta V}\!/_{\delta P}\right)_T \tag{5}$$

It is worth pointing out that other definitions of elastic constant are possible. Elastic modulus *E* measures the resistance to a change in atomic separation distance within the plane of the bond and so can be determined from the linear portion of the interatomic potential. *G* quantifies the resistance to shear loading and *B*, since it corresponds to a volumetric dilatation, is dependent on the electronic properties of a solid, i.e. the compressibility of the electron gas. Elastic moduli are therefore controlled by interatomic interactions and so may be considered a fundamental property of condensed matter. By excitation of longitudinal and transverse phone modes, E and G can, respectively, be calculated if the density (ρ) of the material is known. This is done via an ultrasonic probe which emits and measures the longitudinal (v_l) and transverse (v_t) sound wave velocities, from which E and G can be calculated via Eqs. (6) and (7):

$$E = \rho v_l^2 \tag{6}$$

$$G = \rho v_t^2 \tag{7}$$

E, *G* and *B* can also be calculated from C_{ij} elastic constants. For a material with cubic structure, the number of C_{ij} in the elastic tensor can be reduced from 36 to just 9, due to $C_{ij} = C_{ji}$ and there being strong symmetry in a cubic lattice. The resulting relevant C_{ij} are C_{11}, C_{12} and C_{44}.

$$C_{12} = B + \frac{4G}{3} \tag{8}$$

$$C_{11} = 3B - \frac{C_{11}}{2} \tag{9}$$

$$C_{11} = G \tag{10}$$

$$C' = C_{11} - \frac{C_{12}}{2} \tag{11}$$

The tetragonal shear modulus, *C*, corresponds to a specific phonon vibration mode in the atomic structure, and is thus directional in nature. In comparison, B is non-directional as it relates to a volumetric effect.

$$B = \frac{C_{11} + 2C_{12}}{3} \tag{12}$$

$$G = \frac{C_{11} - C_{12}}{3} \tag{13}$$

3. Elastic and lattice stability criteria

3.1 Lattice stability in perfect crystal

Elastic properties of a material are very important because they check the mechanical stability, ductile or brittle behaviour based on the analysis of elastic constants, C_{ij}, bulk modulus B and shear modulus G. For example, the bulk modulus measures the resistance of the volume variation in a solid and provides an estimation of the elastic response of the materials under hydrostatic pressure. The shear modulus describes the resistance of a material to shape change.

The fundamental understanding of the conditions of mechanical stability of unstressed crystal structure was laid by the work of Max-Born and co-authors in the 1940s [3], and consolidated later in 1954 [3]. This and other text books gave the generic requirements for elastic stability of crystal lattices in terms of elastic constants [3] and offers simplified equivalents of the generic conditions for some high-symmetry classes. The general stability condition can be stated by considering the second-order elastic matrix and the elastic energy of the crystal deformed homogeneously by infinitesimal strain as shown in Eqs. (14) and (15) [3], respectively:

$$C_{ij} = \frac{1}{V_O}\left(\frac{\partial^2 U}{\partial^2 e_i \partial^2 e_j}\right) \tag{14}$$

$$U = \frac{1}{2} V_O \sum_{i,j=1}^{6} C_{ij} e_i e_j + O(e^3) \tag{15}$$

where U is the elastic energy, V_O is the volume of unstressed sample, C_{ij} (I, j = 1–6) is the elastic constant and e_i and e_j are the applied strains [2]. In Eq. (15), O (e3) denotes the terms of numerical error in the order e^3 or higher. A crystal lattice is dynamically said to be stable only if elastic energy U is positive for any small deformation [9], which implies that principal minors of the determinant with elements Cij are all positive [3].

Most real materials (cubic and non-cubic polycrystalline structures) have some types of symmetry, which further reduces the required number of independent elastic moduli. In the case of cubic systems, such as bcc, fcc, NaCl type, or CsCl type) structures, in particular, number of independent elastic moduli is reduced from 36 to 9, as $C_{ij} = C_{ji}$ and there being strong symmetry in the two lattices. Therefore, the conditions for stability reduced to a very simple form using three different elastic constants: C_{11}, C_{22} and C_{44}. The mechanical stability criteria are given by [10]:

$$\begin{gathered} C_{11} - |C_{12},| > 0 \\ C_{11} + 2\,C_{12} > 0 \\ C_{44} > 0 \\ C_{12} > C_{11} \end{gathered} \tag{16}$$

The condition when $B < 0$ is referred to as spinodal instability.

Although hexagonal and tetragonal systems have the same form for the elastic matrix, the hexagonal has five, while tetragonal has six independent elastic constants. By direct calculation of the Eigen values of the stiffness matrix, according to [11], four conditions can be derived for elastic stability in both classes:

$$\begin{gathered} C_{11} > |C_{12},|; 2C_{13}^2 < C_{33}(\,C_{11} + C_{12}) \\ C_{44} > 0; C_{66} > 0 \end{gathered} \tag{17}$$

Similarly, for the orthorhombic system, there are nine independent elastic constants: C_{11}, C_{22}, C_{33}, C_{44}, $C_{12}C_{55}$, C_{66}, C_{23} and C_{13}. The mechanical stability of the structure at each concentration can be judged by calculated elastic stiffness. According to Born's criteria [3], the requirement of mechanical stability in an orthorhombic system leads to the following equations [12].

$$\begin{gathered} C_{11} > 0; C_{22} > 0; C_{44} > 0; C_{33} > 0; C_{55} > 0; C_{66} > 0; \\ C_{11} + C_{22} > 2\,C_{12}; C_{11} + C_{33} > 2\,C_{13}; \\ C_{11} + C_{22}\,C_{33} + 2\,C_{12} + 2\,C_{23} + 2\,C_{13} > 0; \end{gathered} \tag{18}$$

Further to this, conditions for stability for some high-symmetry crystal classes have been studied. However, there is still some confusion about the form of stability criteria for other crystal systems and classes [8].

A crystal lattice is said to be stable in the absence of external load (unstressed condition) and in the harmonic approximation [13] if and only if it has both dynamic and elastic stability. Dynamic stability implies that its phonon modes have positive frequencies for all wave vectors, while its elastic stability is dependent on elastic energy given by Eq. (15) being always positive ($U > 0, \forall \varepsilon \neq 0$). Elastic stability criterion is mathematically equivalent to the following necessary and sufficient conditions: the elastic matrix C is definite exactly positive and all Eigen values of matrix C are positive; all the leading principal and arbitrary minors of matrix C are all positive. The closed form expressions for necessary and sufficient elastic stability criteria for other crystal lattices have been studied. While the stability criterion is linear for some crystal lattices, it is quadratic and even polynomial for others. Thus, the mechanical stability of a crystal is combination of the elastic constant and Born's stability criteria. The elastic constant of a stable crystal must satisfy the Born's criteria to prove its mechanical stability.

3.2 Relative stability of polycrystalline materials

In the case of multi-phase stability, multi-phase composites can be obtained based on multiple scattering theory. For example, polycrystalline materials consisting of two phases, namely cubic and orthorhombic phases can be obtained by homogenising the integral elastic response of the multi-phase polycrystalline samples, following the effective medium approach originally applied by Zeller and Dederichs [13] to determine elastic properties of single-phase polycrystals with cubic symmetry. This type of concept was generalised by Middya and Basu [14] and further extended by Middya [15] and by Raabe et al. [16] to multi-phase composites to determine: (i) the elastic single constants and (ii) the volume fraction of the components within a self-consistent T-matrix solution for the effective medium elastic properties of hexagonal, and orthorhombic polycrystals.

The subset of supercells or cubic and orthorhombic symmetries consisting of three (C_{11}, C_{12}, C_{44}) and nine (C_{11}, C_{12}, C_{13}, C_{22}, C_{23}, C_{33}, C_{44}, C_{55}, C_{66}) elastic constants, respectively, was calculated by employing the methodology explained in [16–18] for the elastic properties of the multi-component alloys. This can be viewed as a macroscopic homogeneous effective medium consisting of microscopic fluctuations and characterised by an effective stiffness of C_{ijkl} defined by:

$$\langle \sigma_{ij}(\boldsymbol{r}) \rangle = C_{ijkl} \langle \epsilon_{kl}(\boldsymbol{r}) \rangle \tag{19}$$

Here, C_{ijkl} is the local elastic constant tensor with $\langle \sigma_{ij}(\boldsymbol{r}) \rangle$ and $\langle \epsilon_{kl}(\boldsymbol{r}) \rangle$ as the local stress and strain field at a point **r**, respectively, and the angular brackets denoting ensemble averages. A repeated index implies the usual summation convention. The effective stiffness of C_{ijkl} is defined by:

$$\langle \sigma_{ij}(\boldsymbol{r}) \rangle = C^{*}_{ijkl} \langle \epsilon_{kl}(\boldsymbol{r}) \rangle \tag{20}$$

Since the aggregate represents a body in equilibrium, $\sigma_{ij}(r)|j = 0,$ where $|j = \partial/\partial r_j$ and the local elastic constant tensor can now be decomposed into an arbitrary constants part (C^{o}_{ijkl}) and a fluctuating part—$\delta C(\boldsymbol{r})$.

$$C_{ijkl}(r) = C^{o}_{ijkl} + \delta C_{ijkl}(r) \tag{21}$$

As shown in [16], an integral part of Eq. (19) is the interactive equivalent solution representing the resulting local strain ϵ distribution (in a short notation) as:

$$\epsilon = \epsilon^0 + GT\epsilon^0, \tag{22}$$

Here, ϵ^0 and G are the strain and modified Green's function of the medium defined by C^O, and the T-matrix given by:

$$T = \delta C(1 - G\delta C)^{-1} \tag{23}$$

Here, ***I*** is equivalent to the unit tensor. Combining Eqs. (21) and (22), we get:

$$C^* = C^0 + \langle T \rangle / (1 + \langle GT \rangle)^{-1} \tag{24}$$

Although Eq. (21) constitutes an exact solution for C^*, finding the exact solution of $\langle T \rangle$ and $\langle GT \rangle$ for realistic cases is impossible. By neglecting the intergranular scattering that may occur in some cases in the form of a grain-to-grain position-orientation correlation function however, the T-matrix can be written in terms of single-grain T-matrix (t_α) for each grain α

$$T \approx \sum_\alpha t_\alpha = \tau. \tag{25}$$

where

$$t_\alpha = \delta C_\alpha + \delta C_\alpha G t_\alpha = \delta C_\alpha (1 - G\delta C_\alpha)^{-1} \tag{26}$$

$$\sum_\alpha \delta C_\alpha = \delta C = C - C^0 \tag{27}$$

Inserting Eq. (21) into (22) leads to:

$$C^* = C^0 + \langle \tau \rangle (1 + \langle G\tau \rangle)^{-1} \tag{28}$$

For single-phase polycrystal, the self-consistent solution of Eq. (11) can be obtained by choosing a C^* that satisfies:

$$\langle \tau \rangle = 0 \tag{29}$$

For a multi-phase polycrystals, a solution to Eq. (4) can be found by evaluating the volume fraction and τ of each phase $i(u^2 \; and \; \tau^2)$, respectively [19], via:

$$\left\langle \sum_i v^2 \tau^2 \right\rangle = 0 \tag{30}$$

The application of the method to both single-phase aggregates and multi-phase composites is relevant to many multi-component alloys. For a single-phase poly-crystal with cubic symmetry [16, 20] to the following expression for B^* and $\mu^* : B^* = B^o$

$$8\mu^{*3} + \left(9B_0 + 4C^{''}\right)\mu^{*2} - 3C_{44}\left(B_O + 4C^{''}\right)\mu^* - 6B_O C_{44} C^{''} = 0 \tag{31}$$

In Eq. (31), three independent single-crystal elastic constants (C_{11}, C_{12}, C_{44}) define the single-crystal bulk modulus $B^o = \left(C_{11}^0 + 2C_{12}^O\right)/3$, the tetragonal shear modulus $C' = (C_{11} - C_{12})/2$ and trigonal shear modulus, C_{44}, $\mu^* = C_{44}^* = \mu^0 + \frac{\langle \tau_{44} \rangle}{1 + G_{44}\tau_{44}}$.

Here, C^*_{44} is the homogenised bulk modulus. The details of the equation for calculating the elastic constants of polycrystals alloy with hexagonal symmetry have been explained elsewhere by [20], and the details here concern polycrystals with orthorhombic symmetry. Eqs. (29) and (30) are reduced to a set of coupled equations for B^* and μ^*:

$$0 = 9(K_V - B^*) + 2\beta(d - c + e) + 3\beta^2\Delta' \tag{32}$$

$$0 = \frac{a - b + \beta(2d - 2c - e) + 3\gamma(d - c + e) + \eta\beta\Delta'}{1 - \alpha\beta - 9\gamma(k_v - B_0) + \beta(\beta + 2\gamma)(c - d) - 2e\beta\gamma - \frac{1}{3}\eta\beta^2\Delta''} + 3\left(\frac{C_{44} - \mu^O}{1 - 2\beta(C_{44} - \mu^O)} + \frac{C_{55} - \mu^O}{1 - 2\beta(C_{55} - \mu^O)} + \frac{C_{66} - \mu^O}{1 - 2\beta(C_{66} - \mu^O)}\right) \tag{33}$$

where

$$9K_v = C_{11} + C_{22} + C_{33} + 2(C_{12} + C_{13} + C_{23}), \tag{34}$$

$$B = 1/3(C_{11} + 2C_{12})\mu^* = C_{44} \tag{35}$$

$$\gamma = 1/9(\eta - 3\beta) \tag{36}$$

$$a = \delta C_{11} + \delta C_{22} + \delta C_{33}; b = \delta C_{12} + \delta C_{13} + \delta C_{23} \tag{37}$$

$$c = \delta C_{11}\delta C_{22} + \delta C_{11}\delta C_{33} + \delta C_{22}\delta C_{33}; d = \delta C_{12}^2 + \delta C_{13}^2 + \delta C_{23}^2 \tag{38}$$

$$e = \delta C_{12}\delta C_{13} + \delta C_{12}\delta C_{23} + \delta C_{13}\delta C_{23} - \delta C_{11}\delta C_{23} - \delta C_{22}\delta C_{13} - \delta C_{33}\delta C_{12} \tag{39}$$

$$\Delta' = \delta C_{11}\delta C_{22}\delta C_{33} + 2\delta C_{12}\delta C_{13}\delta C_{23} - \delta C_{11}\delta C_{23}^2 - \delta C_{22}\delta C_{13}^2 - \delta C_{33}\delta C_{12}^2 \tag{40}$$

$$\delta C_{11} = C_{11} - C_{11}^O = C_{11} - K^0 - \frac{4}{3}\mu^0; \delta C_{22} = C_{22} = C_{22} - K^0 - \frac{4}{3}\mu^0 \tag{41}$$

$$\delta C_{33} = C_{33} - K^0 - \frac{4}{3}\mu^0; \delta C_{12} = C_{12} - C_{12}^O = C_{12} - K^0 + \frac{4}{3}\mu^0 \tag{42}$$

$$\delta C_{13} = C_{13} - K^0 + \frac{4}{3}\mu^0; \delta C_{23} = C_{23} - K^0 + \frac{4}{3}\mu^0 \tag{43}$$

$$\beta = \frac{-3(B^* + 2\mu^*)}{5\mu^*(3B^* + 4\mu^*)}, \tag{44}$$

$$\eta/3 = -1/3B^* + 4\mu^*, \tag{45}$$

$$C_{66} = (1/2)(C_{11} - C_{12}) \tag{46}$$

and orthorhombic symmetry has nine of the single crystal elastic constants, namely: C_{11}, C_{22}, C_{33}, $C_{44}C_{55}$, C_{66}, C_{12}, C_{23} and C_{13}.

The elastic constants of a multi-phase polycrystals were determined directly by coupling Eq. (13) for τ_{44} and the $(\tau_{11} + 2\tau_{12})$ components of the T-matrix. For materials with cubic symmetry, the equation is defined as:

$$5\tau_{44} = \left(\frac{1}{C_{11} - C_{12} - \widetilde{G}^*}^{-\beta}\right)^{-1} + 3\left(\frac{1}{C_{44} - \widetilde{G}^*}^{-2\beta}\right)^{-1} \tag{47}$$

$$\tau_{11} + 2\tau_{12} = \frac{3(C_{11} + 2C_{12}) - 9\widetilde{B}^*}{3 - (C_{11} + 2C_{12}) - 3\widetilde{B}^*} \tag{48}$$

This is where β is defined in Eq. (31) with $\widetilde{G}^*$ and $\widetilde{B}^*$ replacing G^* and B^*. For materials with orthorhombic symmetry, the equation reads:

$$15\tau_{44} = \frac{a-b+\beta(2d-2c-e)+3\gamma(d-c+e)+\upsilon\beta\Delta'}{1-\alpha\beta-9\gamma\left(K_{\upsilon}-\widetilde{B}_{0}\right)+\beta(\beta+2\gamma)(c-d)-2e\beta\gamma-\frac{1}{3}\upsilon\beta^{2}\Delta''}$$
$$+3\left(\frac{C_{44}-\widetilde{G}^{O}}{1-2k\left(C_{44}-\widetilde{G}^{O}\right)}+\frac{C_{55}-\widetilde{G}^{O}}{1-2k\left(C_{55}-\widetilde{G}^{O}\right)}+\frac{C_{66}-\widetilde{\mu}^{O}}{1-2\beta\left(C_{66}-\widetilde{G}^{O}\right)}\right) \quad (49)$$

$$\tau_{11}+2\tau_{12} = \frac{9\left(K_{\upsilon}-\widetilde{B}^{0}\right)+2\beta(d-c+e)+3\beta^{2}\Delta'}{3\left[1-\alpha\beta-9\gamma\left(K_{V}-\widetilde{B}^{0}\right)+\beta(\beta+2\gamma)(c-d)-2e\beta\gamma-\frac{1\eta\beta^{2}\Delta'}{3}\right]} \quad (50)$$

Here, β is defined in Eq. (28), η is defined in Eq. (29), and Δ' in Eq. (23). Here, again G^* and $\widetilde{B}^*$ replaces G^* and B^* in the equations for β, υ and Δ'. As soon as G^* and $\widetilde{B}^*$ have been determined, the homogenised Young's modulus $\left(\widetilde{E}\right)^*$ and Poisson's ratio $(\upsilon)^*$ for (an elastically isotropic) polycrystal can be determined using standard elasticity relationships. The homogenised polycrystalline Young's modulus is calculated using:

$$E^* = \frac{9\widetilde{B}^* G^*}{3\widetilde{B}^* + G^*} \quad (51)$$

$$G^* = \frac{3\widetilde{E}^*\widetilde{B}^*}{9\widetilde{B}^* - \widetilde{E}^*} \quad (52)$$

4. Correlation of elastic constants with properties of polycrystalline materials

In many problem relating to polycrystalline or anisotropic materials, it is customary to make use of the properties in an elastically isotropic materials. Most of the common metals and engineering alloys, however, exhibit a marked degree of anisotropy in their single-crystal elastic behaviour and it is therefore more desirable to obtain same on the bases of anisotropic elastic property. The fundamental factors determining the intrinsic plasticity or brittleness behaviour in solids have great link with interatomic potentials, for instance, there is a correlation with the ratio of the elastic shear modulus μ to the bulk modulus B. It is evident, elastic moduli show trends with a range of properties, including hardness, yield strength, toughness and fragility [21, 22]. In this section, for limitation of space, we will, in particular, consider elastic aspect of polycrystals materials with respect to their dependency on specific crystal structure.

4.1 Elasticity and ductility criteria

Strength and ductility have always been one of the crucial issues to study for metal materials. The tendency of materials to be ductile or brittle is being predicted using models based on elastic constants. Some of these include that of Pugh criterion [23] and Cauchy pressure as defined by Pettifor [24]. Pugh proposed an empirical relationship between the plasticity and fracture properties showing the ratio G/B indicates the intrinsic ability of a crystalline metal to resist fracture and deform plastically [25]. This represents a competition between plasticity and fracture considering that B and G represent resistance to fracture and plastic deformation, respectively. Thus, the force required to propagate a dislocation is proportional to Gb where b is the Burgers vector. This implies that a material with high value of the ratio tends to be brittle (fracture is easier and plasticity is much less), while a low value indicates ductility (plasticity is easier and fracture is not). Fracture strength is also proportional to Ba (a, is the lattice constant) since B is related to surface energy, which indicates brittle fracture strength.

These empirical observations implicate G/B as explaining well brittle or tough behaviour [19, 26]. Pugh's criterion is the most widely used model to predict plastic behaviour of materials [27]. Since yield strength and fracture stress scale with shear modulus and elastic constant, respectively, the Pugh's ratio determines the likelihood of material's failure. If the effect of crystal structure is neglected, high value of Pugh's ratio indicates that a material is prone to brittle failure, while low value of G/B implies ductile failure. The large data on polycrystalline pure metals collected by Pugh [2], when he provided a qualitative ranking from ductile (e.g. Ag, Au, Cd, Cu) to brittle (e.g. Be, Ir) behaviour as G/B increases. For cubic close-packed (ccp) metals, the critical ratio $(G/B)_{crit}$ dividing the two regimes is in the range 0.43–0.56, and for hexagonal close-packed metals, it is 0.60–0.63. Cottrell [28] has estimated $(G/B)_{crit}$ for transgranular fracture from measured surface energies: 0.32–0.57 for ccp metals and 0.35–0.68 for body-centred cubic metals. The spread in values for each structure type largely indicates the interrelationship between crystal structure and elastic constant. Each structure type, however, includes metals with widely differing degrees of elastic anisotropy. Detailed analysis requires knowledge of the relevant elastic constants.

On the other hand, the Cauchy pressure ductility criterion is associated with elastic constants of single cubic crystals such as C_{12}–C_{44} and is useful in describing the nature of bonding in a material [27]. When a material has high resistance to bond bending as found in covalently bonded solids, it will have a negative Cauchy pressure ($C_{44} > C_{12}$). This is in contrast with materials with metallic bonding which exhibit positive Cauchy pressure. When compared with Pugh's ductility criterion, ductile and brittle behaviours are considered to be indicated by a positive and a negative Cauchy pressure, respectively. Although Pugh's and Cauchy pressure criteria are adjured to be based on easily measurable properties of materials such as elastic constants, they do not give the critical value dividing brittle and ductile materials. It is proven in certain materials, including metallic glasses and composites, which religiously respect this dividing line [21]. The behaviour is shown graphically in **Figure 1**. A summary of the correlation between C_{12}–C_{44} and $(G/B)_{crit}$

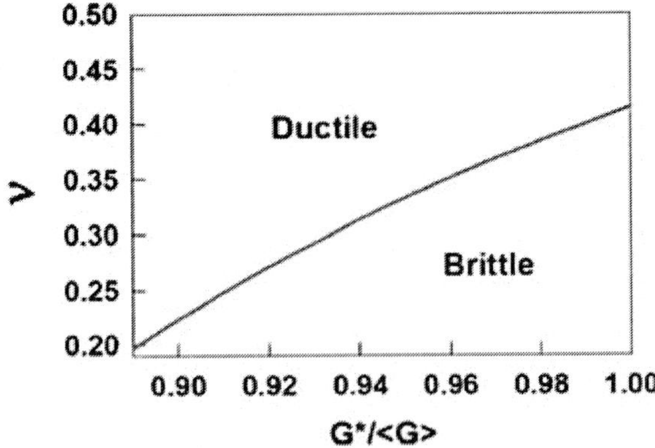

Figure 1.
Ductile and brittle phase fields in metallic glasses, where G is the local modulus and 'G' the global modulus. Decreasing the fraction of low G sites reduces the need for a globally low ν (culled from [28]).*

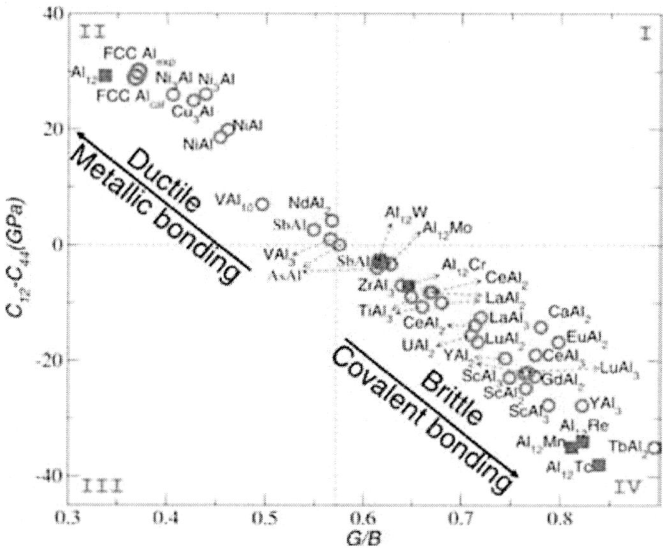

Figure 2.
Correlation between C_{12}–C_{44} and G/B for 35 aluminides (culled from [29]).

for a wide variety of aluminide group of materials is displayed in **Figure 2**. As can be seen, it is evident that an intrinsic correlation between strength and ductility of Al-based materials. It has been observed the criteria indicate a trend in a class of materials with similar deformation mechanism, but is limited by the effects of specimen sizes and crystal structures on deformation processes.

Several authors have studied elastic softening behaviour and recent evidence suggests elastic moduli manifest array of trends with a range of properties including mechanical such as hardness, yield strength, toughness and fragility [22, 30]. In early 1950s, Gilman and Cohen [31] made a historic revelations when they observed that there is a linear correlation between the hardness and elasticity in polycrystalline materials. Nevertheless, successive studies demonstrated that an uniformed linear correlation between hardness and bulk modulus does not really hold for a variety of materials [29] as illustrated in **Figure 3(a)**. Following this, Tester [32] proposed a better empirical link between hardness and shear modulus (G), as illustrated in **Figure 3(b)**. Although, the link between hardness and elastic shear modulus can be arguable, it is certain that he had demonstrated that the shear modulus, the resistance to reversible deformation under shear strain, can correctly provide a key assessment of hardness or ductility criteria for some materials. It is well known that some phase exhibits more hardness or ductility properties than others. Accordingly, it is fair to say that such descriptions could lead to further outlandish discovery in connections with regards to phase components in polycrystalline solids.

4.2 Elastic moduli and martensitic transformation

Martensitic transformation (MT) is a first-order phase type of transformation from a high-symmetry phase (austenite) at high temperature to a crystallographically low-symmetry phase (martensite) at low temperature. Martensitic behaviour has been extensively studied for decades because of its importance

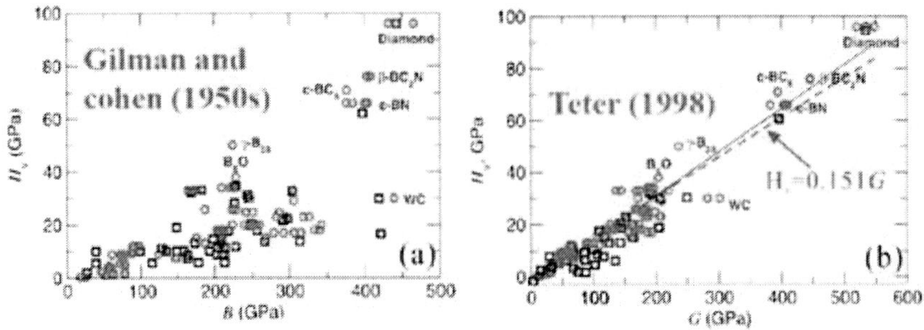

Figure 3.
Correlation of experimental Vickers hardness (H_v) with (a) bulk modulus (B) and with (b) shear modulus (G) for 39 compounds [29].

in metallurgy and its key role in shape memory phenomenon. Shape memory alloys (SMA) are materials such as TiNi and TiNi-based alloys [33], Ti-Nb [34], Ti-Mo [20, 31] etc. that exhibit diffusion-less first-order martensitic phase transitions induced by the change of temperature and/or stress. The relation between softening of elastic constants and martensitic transformation has attracted considerable attention for many years and has been discussed by many researchers [35, 36]. This interesting feature of martensitic transformation in shape memory alloys is the existence of precursor phenomena [1, 2]. The relations between MT temperature and elastic constants were investigated by Ren et al. [36]. Experiments [37] indicate that martensitic transition occurs at almost constant values of C'. Slight change in composition would cause strong deviation in the critical temperature at which C' softens to a critical value and martensitic transition occurs. In some alloys exhibiting martensitic transformation, softening of elastic constants $C' = (C_{11} - C_{12})/2$ and large elastic anisotropy, $A = (C_{44})/C'$ was observed in the parent phase, but the significance of the softening is largely different between the alloys. For example, Earlier Takashi Fukuda and co-workers [34] observed the value of C' near the transformation start temperature is approximately 0.01 GPa in In-27Ti (at %) alloy [38], 1 GPa in Au-30Cu-47Zn (at %) alloy [37], 5 GPa in Fe-30Pd (at %) alloy [39], 8 GPa in Cu-14Al-4Ni (at %) alloy [9], and 14 GPa in Ti-50.8Ni (at %) [33] and Al-63.2Ni (at %) alloys [40]. Because of such a large distribution of C' at the Ms temperature, the influence of softening of C' on martensitic transformation is expected to be significantly different between these alloys. Martensitic transformation in some alloys is probably strongly related to the softening of C', while that in others is weakly related despite the fact that the softening appears before the transformation.

Previously, Zener [5] established a correlation between the magnitudes of $C' = (C_{11} - C_{12})/2$ elastic shear modulus in metallic bcc structures with the occurrence of martensitic phase transformations suggesting links with phase stability, via the atomic interactions. He observed that the large value suggests that C' is much smaller than C_{44} and that MT temperature is dominated by C' [5]. Thus, independent elastic constants are needed to characterize the material response, such as Martensitic transformations (MTs), Shape memory etc. Martensitic transformations (MTs) are often accompanied by elastic modulus softening (acoustic phonon softening) [5]. This explains the strong

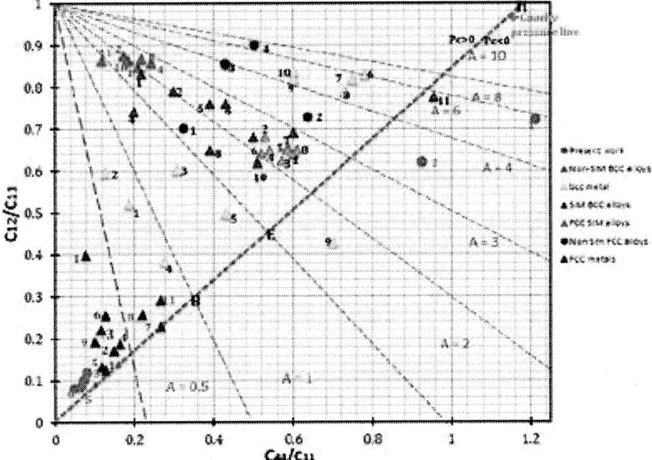

Figure 4.
Correlations between parameters reduced elastic-stiffness coefficients $(C_{12})/(C_{11})$ *vs.* $(C_{44})/(C_{11})$ *for several classes of shape memory materials (culled from [34]).*

composition-dependence of MT temperature. As a result, the modulus softens abruptly within a narrow temperature window around martensitic start temperature, Ms. However, this is unsurprising since it is well known that they are a consequence of weak restoring forces in specific crystallographic directions that announce the possibility of a dynamical instability. The elastic constants are closely related to the acoustic lattice vibrations or even atomic bondings in crystals, and accordingly will be related to the transformation mechanism for not only the Martensitic alloys but also any other compounds which accompany shear-like or displacive transitions.

Following the above, Nnamchi et al. [41] in a recent study considered the link between different groups of shape memory materials with elastic systematics found a clear delineated in a 2D plot of two dimensionless ratios of elastic constants or reduced elastic-stiffness coefficients, $(C_{12})/(C_{11})$ vs. $(C_{44})/(C_{11})$ formally popularised earlier by Blackman [42], It is only one table with different sections. (see **Figure 4** and **Tables 1** and **2**). This reveals among others the elastic anisotropy, proximity to Born mechanical instability, elastic-constants (interatomic-bonding) changes caused by alloying, pressure, temperature, phase transformations and similarities in types of interatomic bonding. The significance of the softening is largely different between the alloys. Inspecting the diagram, we notice materials with similar chemical bonding tend to fall in the same region of the diagrams. Such diagrams provide many uses.

	This work	C_{11}	C_{12}	C_{12}/C_{11}	C_{44}/C_{11}	**Ref.**
1	Ti-3Mo	159.3	115	0.72	1.21	[34]
2	Ti-6Mo	111.3	69.07	0.62	0.93	[34]
3	Ti-10Mo	167	19.6	0.12	0.081	[34]
4	Ti-14Mo	179.2	17.9	0.10	0.074	[34]

	This work	C_{11}	C_{12}	C_{12}/C_{11}	C_{44}/C_{11}	Ref.
5	Ti-18Mo	192.6	16.3	0.085	0.066	[34]
6	Ti-23Mo	197.5	16	0.081	0.051	[34]
	Non-SIM (BCC)alloys					
1	Ti50Ni30Cu20	209	183	0.88	0.17	[43]
2	Ti-50Ni	165	140	0.85	0.21	[44]
3	Ti-29Nb-13Ta-4.6Zr	67.1	39.9	0.87	0.19	[45]
4	Ti-30Nb-10Ta-5Zr	128	92	0.86	0.24	[46]
5	Ti-35Nb	163.5	142	0.87	0.22	[47]
6	Ti-30Nb-5Ta-5Zr	70	30	0.87	0.185	[48]
7	Ti-32.7Nb-11.6Ta-4.49Zr-0.066O-0.052N	137	91.1	0.86	0.12	[49]
	Non-SIM (FCC) alloys					
1	Ag-75Au	230	161.5	0.702	0.33	[43]
2	Cu-4.17Si	117	85.2	0.73	0.64	[43]
3	α-Ag-2.4Zn	190	162	0.85	0.43	[43]
4	α-Cu-9.98Al	199	179	0.89	0.50	[43]
5	α-Cu-22.7Zn	158.9	136.2	0.86	0.43	[43]
	SIM (BCC) alloys					
1	Ti-35.37Nb	130.2	52	0.40	0.078	[50]
2	Ti-35Nb-2Zr-0.7Ta	183	31.4	0.17	0.15	[49]
3	Ti-35.4Nb-1.9Ta-2.8Zr-0.37O	122	27	0.22	0.11	[49]
4	Ti-24.1Nb-4Zr-8.06Sn-0.15O	140	26.3	0.19	0.16	[49]
5	Ti-35Nb-10Ta-4.6Zr-0.16O	102.5	36	0.16	0.12	[51]
6	Ti-23.9Nb-3.75Zr-8.01Sn-0.04O	157.2	36	0.26	0.127	[51]
7	Ti-24Nb-4Zr-7.9Sn-0.17O			0.23	0.22	[51]
8	Ti-24Nb-4Zr-7.6Sn-0.07O	122	31.4	0.26	0.21	[49]
9	Ti-35.2Nb-10.5Ta-4.97Zr-0.091O-0.014N	140	27	0.19	0.1	[51]
10	Ti-23.9Nb-3.8Zr-7.61Sn-0.08O	102.5	26.3	0.12	0.13	[51]
11	Ti-24Nb-4Zr-7.9Sn	157.2	46	0.29	0.27	[52]
	SIM (FCC) alloys					
1	Cu44.9- 50Zn	125	80	0.64	0.6	[43]
2	Au47.5-50Cd	142	96.77	0.68	0.53	[43]
3	Ag45-50Zn	132.8	83.16	0.63	0.57	[43]
4	γ-FeNi	209	183	0.65	0.54	[53]
5	CuAlNi	142.8	93.7	0.66	0.59	[54]
6	B2-NiTi	162	104	0.64	0.52	[55]
7	Cu2.726A11.122Ni 0.152	137	89.2	0.65	0.59	[56]
8	Cu2.742Al1.105Ni0.152	136	81.763	0.65	0.61	[56]

Table 1.
Elastic constant of some bcc and fcc metals and alloys.

	BCC elements	C12/C11	C44/C114	Ref.
1	V	0.52	0.19	[53]
2	Nb	0.59	0.13	[53]
3	Ta	0.60	0.31	[53]
4	Mo	0.38	0.28	[53]
5	W	0.5	0.43	[53]
6	Li	0.83	0.78	[53]
7	Na	0.82	0.75	[53]
8	K	0.79	0.73	[53]
9	Ba	0.43	0.7	[53]
	FCC elements			[53]
1	Au	0.83	0.22	[53]
2	Pd	0.79	0.3	[53]
3	Pt	0.74	0.2	[53]
4	Ag	0.76	0.39	[53]
5	Cu	0.76	0.43	[53]
6	β-Co	0.69	0.6	[53]
7	α-Sr	0.65	0.39	[53]
8	γ-Fe	0.68	0.5	[53]
9	Ni	0.62	0.51	[53]
10	δ-Pu	0.78	0.96	[53]

Table 2.
Elastic constant of some bcc and fcc metals and alloys.

5. Summary and future challenges

The following bullet points summarise some of the main challenges facing the community.

- Some empirical elastic relationship such as a low G/B ratio (or high ν) favours toughness but also indicates a fragility in polycrystalline materials, though they can be typically difficult to vitrify in some polycrystalline materials.

- Some empirical correlations exist in most of the metallic elements in the periodic table have been found, and alloy development has moved beyond the bucket chemistry type approach used in the early days of elastic properties research. While a number of general guidelines exist for explaining elastic systematics property formation (such as Zener, and Burger's rules), Pugh and Pettifor's criterion [16, 17] in addition to Blackmans have gone beyond simply stating the chemical species that should be present, and their rough proportions, and instead gives exact elastic relationship. However, a more rigorous that delineated the phase stability using systematics could be envisaged in new future.

Glossary of symbols

Symbols used symbols derived from disambiguation (e.g. d for

C_{ijkl}	the local elastic constant tensor with $\langle \sigma_{ij}(\boldsymbol{r}) \rangle$ and $\langle \epsilon_{kl}(\boldsymbol{r}) \rangle$ as the local stress and strain field at a point **r**, respectively, and the angular brackets denote ensemble averages
C_{44}	single crystal bulk modulus; $B^o = \left(C_{11}^0 + 2C_{12}^O\right)/3$
$C' = (C_{11} - C_{12})/2$	tetragonal shear modulus
$C_{44}, \mu^* = C_{44}^* = \mu^0 + \frac{\langle \tau_{44} \rangle}{1+G_{44}\tau_{44}}$	trigonal shear modulus
G	the ratio of shearing stress τ to shearing strain γ within the proportional limit of a material
B	bulk modulus, ratio between the fluid pressure and the Volumetric Strain
E	modulus of elasticity or Young's modulus
G	modulus of rigidity or shear modulus
V_L and V_S	the ultrasonic longitudinal and shear wave velocities respectively
ρ	the density of the material
$A = (C_{44})/C'$	elastic anisotropy
U	the energy of the crystal, and quadratic function of the strains
V_0	equilibrium volume
e	an elastic strain
$\langle \sigma_{ij}(\boldsymbol{r}) \rangle$	effective stiffness of C_{ijkl}
$\epsilon = \epsilon^0 + GT\epsilon^0$	ϵ^0 and GT are the strain and modify Green's function
T	T-matrix is given by $T = \delta C(1 - G\delta C)^{-1}$
$\boldsymbol{I}$	equivalent to the unit tensor
$\widetilde{Y}*$	the homogenised polycrystalline Young's modulus
$\widetilde{\mu}*$	homogenised polycrystalline Poisson's ratio
M_S	martensite formation start temperature
M_F	martensite finish temperature
SME	shape memory effect

References

[1] A History of the Theory of Elasticity and of the Strength of Materials Vol. 1 (1886) Vol. 2 (1893)

[2] Born M. Mathematical Proceedings of the Cambridge Philosophical Society. 1940;**36**:160

[3] Born M, Huang K. Dynamics Theory of Crystal Lattices. Amen House, London: Oxford University Press; 1954

[4] Nye JF. Physical Properties of Crystals. Oxford, UK: Oxford University Press; 1985

[5] Zener JC. Elastic and Anelastic of Metals. Chicago, Illinois, USA: University of Chicago press; 1948

[6] Pagnotta L. Recent progress in identification methods for the elastic characterization of materials. International Journal of Mechanics. 2008;**2**(4):129-140

[7] Pokluda J et al. Ab initio calculations of mechanical properties: Methods and applications. Progress in Materials Science. 2015;**73**:127-158

[8] Nye JF. Physical Properties of Crystals: Their Representation by Tensors and Matrices. Amen House, London: Oxford University Press; 1985

[9] Grimvall G, Magyari-Kope B, Ozolin V, Persson KA. Lattice instabilities in metallic elements. Reviews of Modern Physics. 2012;**84**:945-986

[10] Mouhat F, Coudert F-X. Necessary and sufficient elastic stability conditions in various crystal systems. Physiological Reviews. 2014;**B90**:224104

[11] Milman V, Warren MC. Journal of Physics: Condensed Matter. 2001;**13**:241

[12] Rajamallu K, Niranjan MK, Dey SR. Materials Science and Engineering: C. 2015;**50**:52-58

[13] Zeller R, Dederichs PH. Physica Status Solidi B. 1973;**55**:831-842

[14] Middya TR, Basu AN. Journal of Applied Physics. 1986;**59**:2368-2375

[15] Middya TR, Paul M, Basu AN. Journal of Applied Physics. 1986;**59**: 2376-2381

[16] Raabe D, Sander B, Friak M, Ma D, Neugebauer J. Acta Materialia. 2007;**55**: 4475-4487

[17] Eriksson O. Encyclopaedia of Materials: Science and Technology. Amsterdam: Elsevier; 2003. pp. 1-11

[18] Beznosikov BV. Journal of Structural Chemistry. 2003;**44**:885

[19] Thompson JR, Rice JR. Philosophical Magazine Letters. 1974;**29**:269-277

[20] Nnamchi PS. Materials and Design. 2016;**108**:60-67

[21] Wang WH. Journal of Non-Crystalline Solids. 2005;**351**:1481-1485

[22] Lewandowski JJ, Wang WH, Greer AL. Philosophical Magazine Letters. 2005;**85**:77-87

[23] Pugh SF. XCII. Relations between the elastic moduli and the plastic properties of polycrystalline pure metals. The London, Edinburgh, and Dublin Philosophical Magazine and Journal of Science. 1954;**45**:823-843

[24] Pettifor DG. Theoretical predictions of structure and related properties of intermetallics. Materials Science and Technology. 1992;**8**:345-349

[25] Pugh SF. Philosophical Magazine. 1954;**45**:834

[26] Thompson RP, Clegg WP. Predicting whether a material is ductile

or brittle. Current Opinion in Solid State & Materials Science. 2018;**22**(3):100-108. DOI: 10.1016/j.cossms.2018.04.001

[27] Cottrell AH. In: Charles JA, Smith GC, editors. Advances in Physical Metallurgy. London: Institute of Metals; 1990. pp. 181-187

[28] Poon SJ, Zhu A, Shiflet GJ. Applied Physics Letters. 2008;**92**:261902

[29] Tester DM. MRS Bulletin. 1998; **23**:22

[30] Gilman JJ, Cohen. Science. 1993; **261**:1436-1439

[31] Gao FM, Gao LH. Journal of Superhard Materials. 2010;**32**:148

[32] Ren X. Materials Science and Engineering. 2001;**A312**:196-206

[33] Fukuda T, Kakeshita T. Metals. 2017;7:156. DOI: 10.3390/met7050156

[34] Nnamchi PS, Todd I, Rainforth MW. Novel approach to Ti alloy design for biomedical applications. UK: Department of Materials Science and Engineering, University of Sheffield; 2014

[35] Murakami Y. Lattice softening, phase stability and elastic anomaly of the β-Au-Cu-Zn alloys. Journal of the Physical Society of Japan. 1972;**33**: 1350-1360

[36] Gunton DJ, Sanuders GA. The elastic behavior of In-Tl alloys in the vicinity of the martensitic transformation. Solid State Communications. 1974;**14**:865-868

[37] Michal L et al. Materials Science and Engineering A. 2007;**462**:320-324

[38] Scripta Materialia. 1998;**38**(11): 1669-1675

[39] Muto S, Oshima R, Fujita FE. Elastic softening and elastic strain energy consideration in the fcc-fct transformation of FePd alloys. Acta Metallurgica et Materialia. 1990;**38**: 685-694

[40] Enami K, Hasunuma J, Nagasawa A, Nenno S. Elastic softening and electron-diffraction anomalies prior to the martensitic transformation in Ni-Al β1 alloy. Scripta Metallurgica. 1976;**01**: 879-884

[41] Nakanishi N. Elastic constants as they relate to lattice properties and martensite formation. Progress in Materials Science. 1980;**24**:143-265

[42] Blackman M. Proceedings. Royal Society of London. 1938;**164**:62

[43] Ren X et al. Philosophical Magazine A. 1999;**79**:31-41

[44] Hatcher N, Kontsevo Yu O, Freeman A. Physical Review B. 2009;**80**: 1-18

[45] Li SJ, Cui TC, Hao YL, Yang R. Acta Biomaterialia. 2008;**4**:305-317

[46] Nobuhito S, Niinomi M, Akahori T. Materials Transactions. 2004;**45**(4): 1113-1119

[47] Zhengjie L, Wan L, Xiaobing X, Weijie L, Qina J, Zhang D. Materials Science and Engineering: C. 2013;**33**: 4551-4561

[48] Hou FQ, Li SJ, Hao YL, Yang R. Scripta Materialia. 2010;**63**:54-57

[49] Tane M, Nakano T, Kuramoto S, Hara M, Niinomi M, Takesue N, et al. Acta Materialia. 2011;**59**:6975-6988

[50] Kim HY, Ikehara Y, Kim JI, Hosoda H, Miyaaki S. Acta Metallurgica. 2006; **54**:2419

[51] Talling RJ, Dashwood RJ, Jackson M, Kuramoto S, Dye D. Scripta Materialia. 2008;**59**:669

[52] Hao Y, Li S, Sun B, Sui M, Yang R. Physical Review Letters. 2008;**98**:1-4

[53] Paszkiewicz, Wolski. Journal de Physique(Conference series). 2008;**104**: 012038

[54] Mañosa L et al. Physical Review B. 1994;**49**:9969-9972

[55] Zhou L, Cornely P, Trivisonno J, Lahrman D. An ultrasonic study of the martensitic phase transformation in NiAl alloys. Honolulu, HI, USA: IEEE Symposium on Ultrasonics; 1990; **3**:1389-1329

[56] Sedlak P, Seiner H, Landa M, Novak V, Sittner P, Li M. Acta Materialia. 2005; **53**:3643-3661

Chapter 5

Repair Inspection Technique Based on Elastic-Wave Tomography Applied for Deteriorated Concrete Structures

Abstract

Applying elastic wave tomography as an innovative NDT method, the evaluation of velocity distribution in three-dimensional (3D) before and after the repair is introduced in this study. The increase in the velocity with penetration of the repair material according to the repair effect is identified visually and quantitatively. The 3D tomography technique is newly proposed for one-side access inspection, using drill hammering to generate an elastic wave. Accordingly, the elastic wave velocity distribution result enables to visualize the internal quality of concrete after patch repair is successfully done. In addition, an attempt for reinforced concrete (RC) slab panels is made to confirm the effectiveness of the repair by comparing the velocity distribution of elastic waves obtained from acoustic emission (AE) tomography analysis, before and after the repair. Thus, the velocity recoveries due to injection are found in all the slab panels, and it is confirmed that the elastic wave velocities obtained using this technique can serve as an indicator for examining the state of crack and void filling with injected material. Further, a good correlation is found between the low-velocity region before repair and the amount of injection. These results show the potential of the AE tomography technique to be used as a method for estimating the effect of injection repair.

Keywords: elastic wave, acoustic emission, wave velocity distribution, tomography, repair method

1. Introduction

It is highly demanded to establish sufficient management systems for the inspection of existing concrete infrastructures in order to manage and extend their service lives. As for aging infrastructure, severe deterioration is currently reported, where it is known as a critical issue in our society, and large budgets are required to repair damaged structures. Since budgetary restrictions are often imposed, preventive and proactive maintenance techniques of infrastructure are sufficiently needed with nondestructive testing (NDT) methods. In addition to conventional NDT, innovative methods must be established to appropriately assess and evaluate damage and repair and retrofit recovery in concrete structures. Inspection techniques after crack repair methods application for existing structures to assess repair installations have not yet been practically developed, meanwhile improper repair efforts have resulted in re-deterioration. Refilling internal cracks with repair materials from the concrete surface, epoxy injection, and patch repair methods are widely implemented. In most cases, re-deterioration could be led by the unknown and remained internal defects. Consequently, it is very important to implement and establish inspection techniques which can visualize internal defects as a countermeasure with repair works.

For such infrastructure as bridges and tunnels, it is generally recognized that appropriate maintenance works are necessary. Prior to extensive damage and failure in existing structures, essential issues include establishing a maintenance system for reinforced concrete (RC) members with the sufficient measures. Epoxy injection and patch repair methods have been widely and practically introduced to repair and re-strengthen RC members. However, insufficient repair works are unfortunately often reported, and these works have potentially resulted in re-deterioration because more improvement is needed for inspection techniques to estimate the quality of repair and recovery.

Developing nondestructive testing and evaluation methods is strongly demanded for concrete structures to quantify or assure the repair and retrofit recovery. The International Union of Laboratories and Experts in Construction Materials, Systems, and Structures (RILEM) launched a technical committee on innovative NDT for repair and retrofit recovery [1]. Tomography techniques are studied based on elastic wave and acoustic emission (AE) to visualize, internal defects in three-dimension concrete with the committee's activities. These techniques applicability has already been published in terms of elastic wave tomography [2, 3] and AE tomography [4, 5].

Using parameters of elastic wave such as amplitudes and elastic wave velocities, internal distributions are obtained by the tomography technique. Elastic wave velocity is specifically used as the parameter in this study. Both the location of the excitation and the excitation time are known in the mentioned elastic wave tomography. On the other hand, they are unknown for AE tomography. The elastic wave velocity in each set-element over the structure can be calculated. Elastic wave velocity is theoretically associated with elastic modulus of material. The values would vary as low-velocity zones with the presence of such internal defects as cracks and voids.

In a theory of elastic wave propagation inside media, the waves are reflected, diffracted, and scattered where it has voids and cracks. Elastic wave velocity is known to be decreased by the phase divergence. The zones of lower elastic wave velocity corresponding to those of heavier deterioration can be reasonable assumed. The distribution of wave velocities can be accordingly referred to as a good indicator of the internal condition of a concrete structure. Moreover, in order to guarantee whether the injected material is properly filled into cracks by using the crack injection method, the velocity distributions of elastic waves in the applicable regions of RC structures are estimated, before and after the repair, by employing AE tomography method [6].

The repair effects in concrete were evaluated with 3D elastic wave tomography in the present study by means of innovative NDT, which can visually identify the outcome from the repair condition provided by the epoxy injection and patch repair methods. 3D tomography was employed for a 50-year-old concrete pier, which was repaired by epoxy injection method, as well as to a 53-year-old concrete wall, which was repaired by the patch repair method. And, AE tomography was applied to a 46-year-old RC slabs, in which epoxy-based resin was used as the injected material to repair the internal cracks.

As described here, although the epoxy injection and patch repair methods are major repair methods even without the corrosion of the reinforcing bars, there are many reports indicating re-deterioration with insufficient repairs. This study aims to validate the 3D elastic wave tomography and AE tomography technique for inspection of the internal quality of concrete after repair.

2. Examples of existing concrete structures for repair inspection

2.1 Pier

Concrete pier specimen, 600 mm width, 1200 mm height, and 300 mm thickness, is shown in **Figure 1**. About 93 components of syringe-type caulking guns were set into pots for injection and 50 kHz resonance AE sensors were arrayed to receive elastic waves before the injection and 7 days after injection, which is corresponding to the epoxy resin hardening period.

Attached AE sensors to four sides of the pier, as shown in **Figures 2** and **12** sensors were arranged on sides A and B in a 600 × 1200 mm area and 4 sensors were installed on the other sides. About 25 mm diameter steel ball was used for the excitation of elastic wave. In order to identify the impact excitation time, at the closest sensor location, each excitation point was selected.

2.2 Wall

Figure 3 shows concrete wall, 600 × 600 mm, where the patch repair method was applied, following V-shaped concrete removal was conducted for 80 mm depth and 120 mm width since surface cracks with water leakage were observed on the

Figure 1.
Overview of concrete pier.

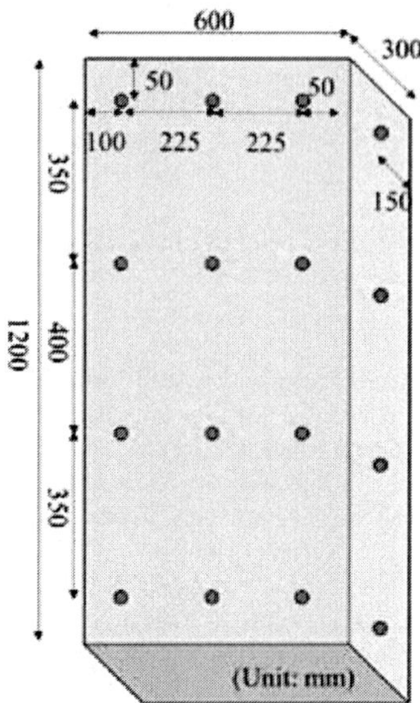

Figure 2.
Sensor arrangement.

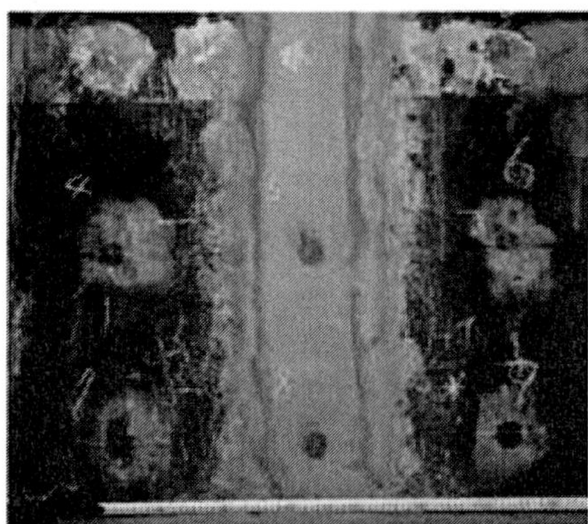

Figure 3.
Overview of concrete wall.

surface of the tunnel-lining concrete. Then, polymer cement mortar with a water-to-cement ratio W/C = 25% was used to fill the crack. Employing micro-core drilling and hammering as one-sided access measurement, the wave signals generated inside the concrete were detected. A 12 mm diameter micro-coring was performed up to 200 mm depth. A curved edge 6 mm diameter steel bar was inserted into the bit hole. The head of steel bar was hit by 25 mm diameter spherical steel ball. Hammering the steel bar without touching the hole wall, elastic waves could only be generated at the hole end in the depth direction. About 60 kHz resonance AE sensors were installed to detect the elastic waves. The sensor arrangements and excitation points are shown in **Figure 4**.

2.3 Slab

Figure 5 and **Table 1** show a top view of an RC bridge, and specifications for the measured deck panels. This bridge is a municipal road bridge located in the Hokuriku region, Japan and it has been in service in the last 46 years. Three panels

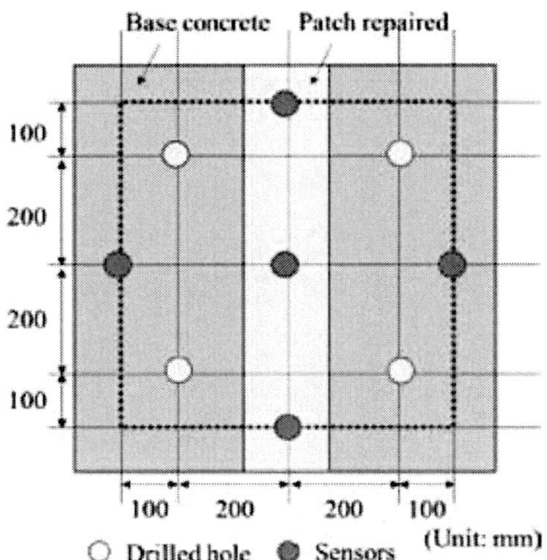

Figure 4.
Locations of drilling and sensor arrangement.

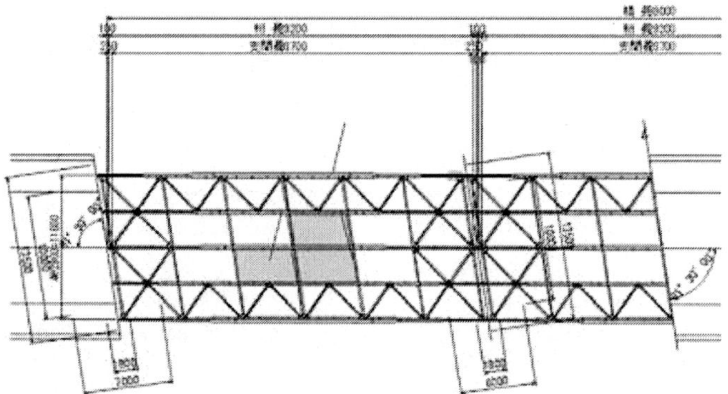

Figure 5.
A top side view of subject bridge.

Type	RC bridge (3 span composite girder bridge)
Length	88.0 m
Age	46 years
Thickness	Slab: 250 mm and asphalt: 50 mm
Condition	Web-shaped cracks were sporadically evident on the concrete surface.

Table 1.
Bridge specifications.

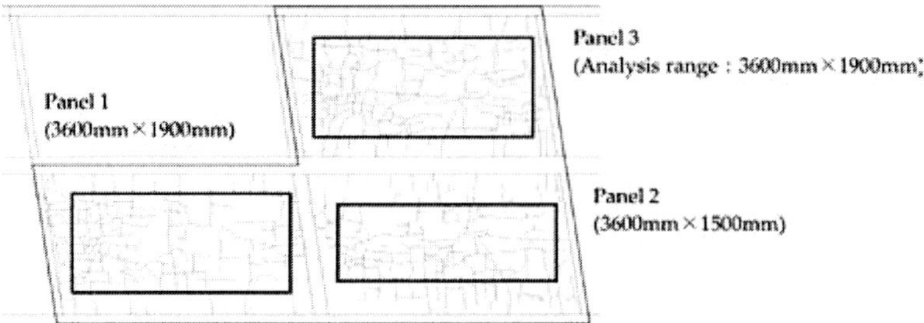

Figure 6.
Sketch of cracking.

highlighted in the figure are selected for the measurement. On all of the slab panels, web-shaped cracks were sporadically evident on the concrete surface. These cracks are thought to be caused primarily by the alkali-silica reaction in concrete. **Figure 6** shows a sketch of cracks obtained through visual inspection from the bottom side of the slab. This figure also shows the area of the tomography analysis for obtaining the velocity distribution. Crack widths are not indicated in figure, but in all the slab panels, the cracks width was smaller than 0.2 mm, and over almost the entire range, the widths were in the range of 0.10–0.15 mm.

3. Data analysis

3.1 Wave detection and computation of elastic wave arrival time

In order to determine the velocity distributions by tomography, the following analytical steps are taken.

First, the arrival time at each sensor was determined with an Akaike Information Criterion (AIC) picker [7, 8]. For the digitized wave record x_k of length N, the AIC value is defined as

$$\mathrm{AIC}(k) = k \times \log\{\mathrm{var}(x[1,k])\} + (N - k - 1) * \log\{\mathrm{var}(x[k+1,N])\} \tag{1}$$

where $\mathrm{var}(x[1,k])$ indicates the variance between x_1 and x_k, and $\mathrm{var}(x[k,N])$ is the variance between x_k and x_N.

The point where AIC value minimizes, applying the least-square method, corresponds to the most suitable separation point of two series of stationary time, the arrival time as the phase onset is thus reasonably determined by the AIC picker. Lower AIC values suggest noise and higher AIC values show the arrival of wave signals. Following the determination of arrival time, the elastic wave velocity is calculated. The observed time of wave propagation T_{obs} is obtained by [9].

$$T_{obs} = T_o - T_s \tag{2}$$

where T_s is the time of excitation and T_o is the arrival time.

3.2 Elastic wave tomography

The reciprocal of the velocity is referred in the elastic wave tomography algorithm to as the "slowness." As shown in **Figure 7**, slowness as the initial parameter

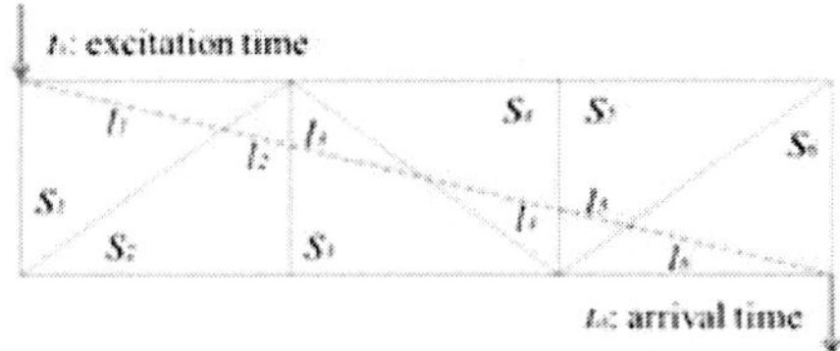

Figure 7.
Slowness for calculation of propagation time.

is provided into each element. Travel time of elastic wave can be computed as elastic velocity is constant in individual element on this ray path. The total of the propagation time calculated by the slowness and the distance in each element (refer to Eq. (3)) derives the propagation time T_{cal}. The difference between the observed propagation time (T_{obs}) and the theoretical propagation time (T_{cal}) is obtained by Eq. (4).

where l_j is the length crossing each element and s_j is the slowness of each element.

$$T_{cal} = \sum_j sj{*}lj \tag{3}$$

$$\Delta T = T_{obs} - T_{cal} \tag{4}$$

s_i is slowness of element i, l_i is length of the ray path in element i. Thus, it is revealed that l_i is essential for the calculation of the travel time.

In order to reduce the difference between the observed propagation time and the theoretical propagation time, the slowness in each element is re-calculated and renewed. The total slowness correction is determined by Eq. (5) and the revised slowness is consequently calculated by Eq. (6).

$$\begin{bmatrix} \Delta s1 \\ \Delta s2 \\ \vdots \\ \Delta sj \end{bmatrix} = \begin{bmatrix} \sum_i \dfrac{\Delta T_i{}^{*} l_{i1}}{L_i} \Big/ \sum_i l_{i1} \\ \sum_i \dfrac{\Delta T_i{}^{*} l_{i2}}{L_i} \Big/ \sum_i l_{i2} \\ \vdots \\ \sum_i \dfrac{\Delta T_i{}^{*} l_{ij}}{L_i} \Big/ \sum_i l_{ij} \end{bmatrix} \tag{5}$$

$$s'j = sj + \Delta sj \tag{6}$$

where L_i is the total distance of wave propagation through the i-element.

Proceeding the iteration based on Eqs. (5) and (6) as shown in **Figure 8**, the optimal slowness, eventually the velocity, in each element corresponding to the observed propagation times of multiple paths over the interested area is determined as well as the velocity distribution.

In order to determine the ray path more accurately, the ray trace algorithm is applied, taking into account detours of elastic waves due to the reflection and diffraction. Following 3D ray trace algorithm, which was proposed in previous research [3], the arrival time of each wave is obtained. Correction of the slowness in each element is carried out according to the error between the observed first travel time and computed value in the element, using 3D finite elements for meshing of target space in the present algorithm. Wave velocities between 2000 and 4500 m/s are given for the tomography results as the range of wave velocities in concrete.

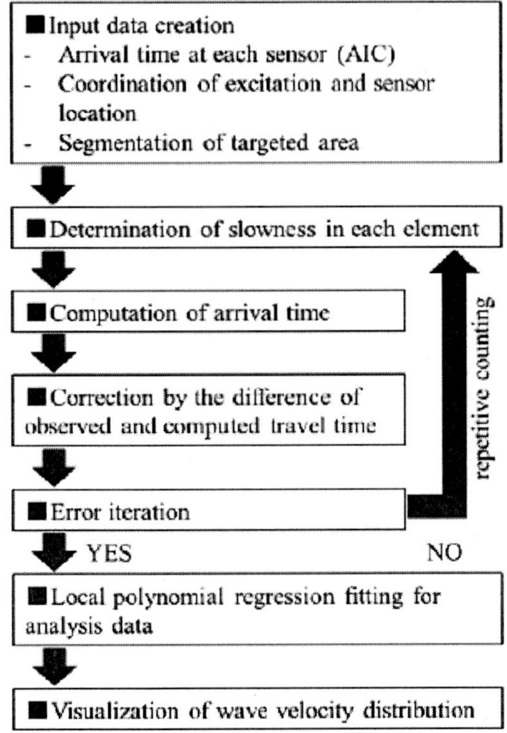

Figure 8.
Analytical procedure for 3D tomography.

3.3 Acoustic emission (AE) tomography

AE tomography is a method for obtaining a velocity distribution by finding the travel time from an AE source to each sensor. Thus, it is necessary to obtain the position of the transmission source as accurately as possible. With the conventional ranging technique, which assumes that the propagation velocity is fixed, considerable errors are expected in the case that the tomography technique is applied to such a heterogeneous material as concrete. Consequently, a new ranging technique incorporating with the ray tracing concept has been developed as a pre-processing technique for AE tomography [6]. The ranging technique using ray tracing is illustrated in **Figure 9**. As shown in the diagram, ray tracing is performed from the received point j to all other nodes i, and the theoretical travel time T_{ji} to each node is calculated. The shortest transmission time is determined from the differences between T_{ji} and the initial travel time T_j at the received point j. The procedure is repeated for the number of received points N, and finally the node, where the variance of estimated arrival times estimated from Eqs. (7) and (8) becomes the minimal, is taken to be the transmission point. In Eq. (7), T_{mi} is the mean value of the estimated transmission times at each node i, and in Eq. (8), σ_i is the variance of the estimated transmission times at each node i.

$$T_{mi} = \frac{\sum_j (T_{ji} - T_j)}{N} \tag{7}$$

$$\sigma_i = \frac{\sum_j \left(T_{ji} - T_j - T_{mi}\right)^2}{N} \tag{8}$$

3.4 Excitation method of elastic waves and analysis model

Figure 10 shows the model of AE tomography analysis and the positions of receiving sensors. The shaded part at the top of the model indicates the asphalt layer (thickness: 50 mm). Analyzed regions for slab panels 1 and 3 were set to be 3600 × 1900 mm. Concerning slab panel 2, there were limitations on the sensor positions, and thus the region was set to be 3600 × 1500 mm. As elements for AE tomography analysis, the applicable region was divided by 16 × 8 in total of 128 elements. In AE tomography, elastic waves were excited by the steel ball drop. A steel ball of 5 mm diameter was dropped at several locations for 12 minutes from the asphalt surface, consciously ensuring that the distribution of impact points was as uniform as possible at the target area. The steel ball dropping is illustrated in

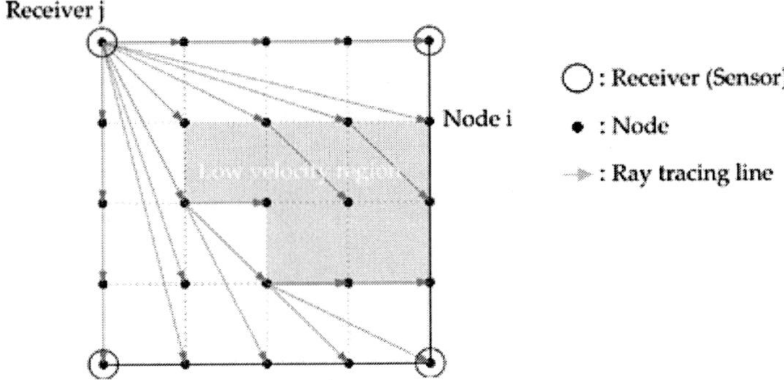

Figure 9.
Overview of transmission source estimation using ray tracing.

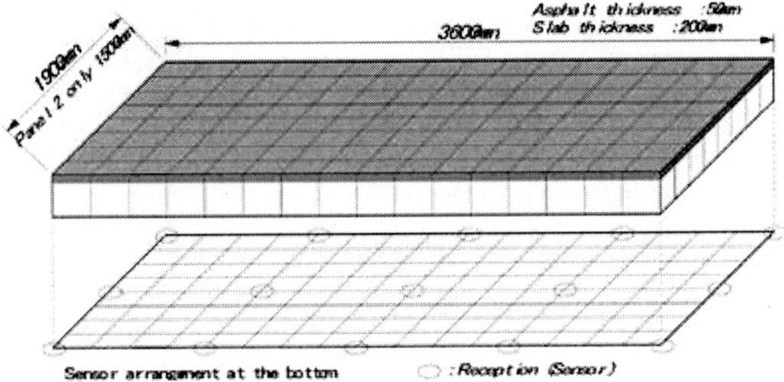

Figure 10.
Analysis model for AE tomography.

Figure 11.
Steel ball dropping.

Figure 11. In AE tomography, the measurements were performed using an acceleration measurement system (TEAC). About 15 piezoelectric accelerometers with the frequency response from 3 Hz to 15 kHz were employed as receiving sensors. The point at which AIC is the minimum is determined as the arrival time of the wave. However, when the S/N ratio is low, it is difficult to identify the minimum value of AIC. Thus, a reliability parameter is developed for reading the initial travel time. The index is proposed as a measure for the identification of the rising edge of the wave [10]. It is found that readings of the initial travel times reasonably converge if the index is 0.05 or higher. In the present chapter, elastic waves with the index of 0.1 or higher are analyzed. AIC (k_{min}) indicates the minimum value of AIC, that is, corresponding to the initial travel time.

4. Results and discussion

4.1 Pire

4.1.1 Wave detection and computation of elastic wave arrival time

In order to investigate the epoxy-injected situation in damaged concrete, black light (ultraviolet light) was irradiated on the cored sample so that the injected material (epoxy resin) was colored in blue as shown in **Figure 12**. It is confirmed that epoxy resin was successfully injected into the concrete cover (up to 10 cm) and over the depth of the reinforcing bars (from 10 to 15 cm). The injected material can penetrate cracks even smaller than 0.1 mm in width [11].

4.1.2 3D elastic wave tomography

Figure 12 shows elastic wave tomography results before and after epoxy injection. The overview of injection repair method is shown in **Figure 13**. The wave velocities after the repair indicate clearly higher values than those before. The velocities are mostly higher than approximately 3000 m/s. This result implies that the epoxy injected from the surface of the pier could be filled and hardened sufficiently inside the media via cracks. However, at the central portions of the concrete pier, wave velocities are still lower than 2500 m/s, namely, the epoxy injection only guarantees the shallow zone repair from the concrete surface. The velocity

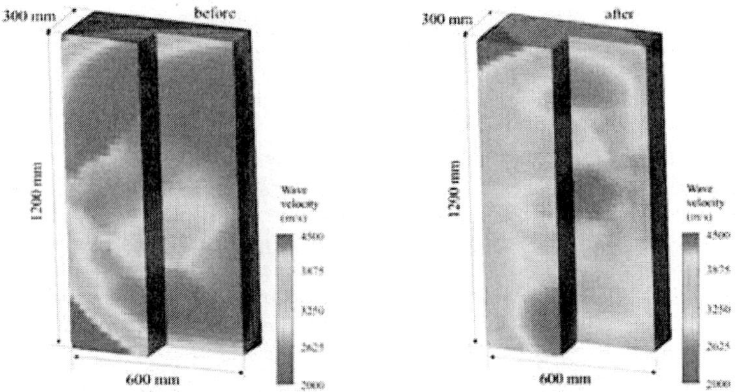

Figure 12.
Results of elastic wave tomography in 3D (left: before injection and right: after injection).

Figure 13.
Repair by epoxy injection.

distribution given by 3D elastic wave tomography shows the conditions inside the concrete, in particular, whether the epoxy is fully penetrated into the interior, while it is noted that the tomography technique could assess the repair level, which is not visually clarified on the exterior.

Figure 14 shows wave velocity histogram before and after the repair consequences. The mean value after the repair is higher than that before, and the variation decreases. Since the velocities lower than 2500 m/s are rarely observed in the histogram, concrete of the pier is repaired after the injection.

4.1.3 Amount of epoxy injection and alteration of wave velocities

Figures 15 and **16** respectively show the tomography results the injected epoxy amount at Side A (referred as the front surface in **Figure 12**) and those at Side B (referred as the back surface in **Figure 12**). They are only the tomograms of wave velocities at the surface layer, comparing with the amount of epoxy injection. The injection pots which added syringe refilling are colored in red because the caulking guns were replaced and refilled with epoxy until the spring-loaded gun automatically stopped the injection.

The velocity distribution alteration reasonably correlates with the epoxy injection amount. Concrete property improvement suggested by the velocity recovery is also roughly confirmed with the epoxy injection amount.

Velocity-improved areas in **Figures 15** and **16** are relatively observed in the areas, where additional injection was installed because of their porous media due to heavy deterioration (red colored in **Figures 15(d)** and **16(d)**). Less improvement of the velocity are observed even after the repair at the bottom-right corner of side A (see **Figure 15**), where the method did not enable to penetrate the resin sufficiently into the concrete because of their lower connectivity of internal cracks.

4.2 Wall

3D elastic wave tomography technique mentioned above was challengingly applied to confirm patch repair effect for concrete wall of an existing structure. In this study, the technique was introduced as a method to evaluate the retrofit recovery. There is currently no NDT technique applicable in terms of in-situ measurement.

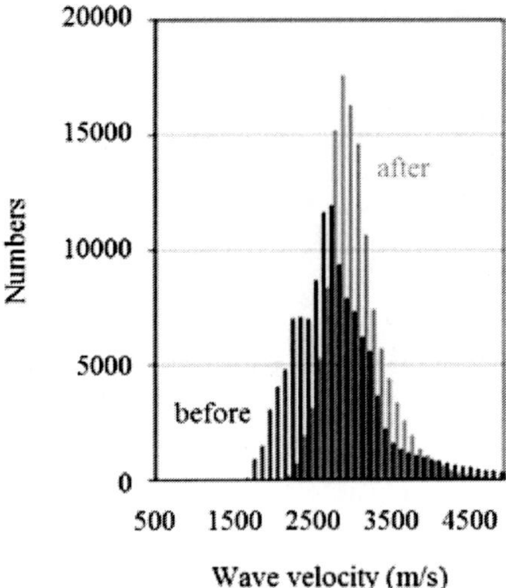

Figure 14.
Histogram of wave velocities.

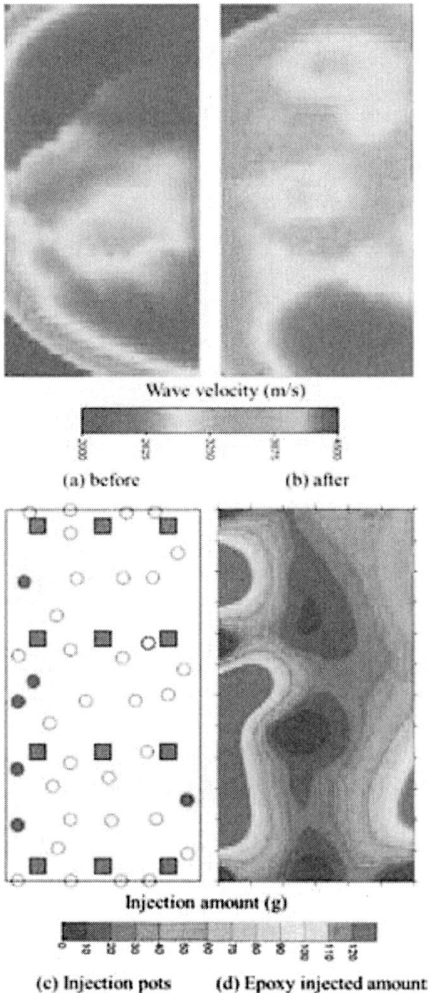

Figure 15.
Results of wave velocities and amount of epoxy injection (Side A).

4.2.1 Procedures of drilling and excitation

Introducing micro-core drilling, excitation of elastic waves was driven. The technique is proposed and applied usefully for one-side access inspection works. **Figure 17** shows the illustration of test procedure schematically.

On the surface of concrete wall with a surface crack (a), a V-cut concrete removal is performed (b), followed by a patch repair method with polymer cement mortar grouting (c), 12 mm diameter bit hole of 200 mm depth is drilled by micro-coring (d), at each concrete surface point. With the sensor array on the surface (e), 6 mm diameter a steel bar is inserted into the hole and the steel bar head is hit by a 25 mm diameter steel sphere ball.

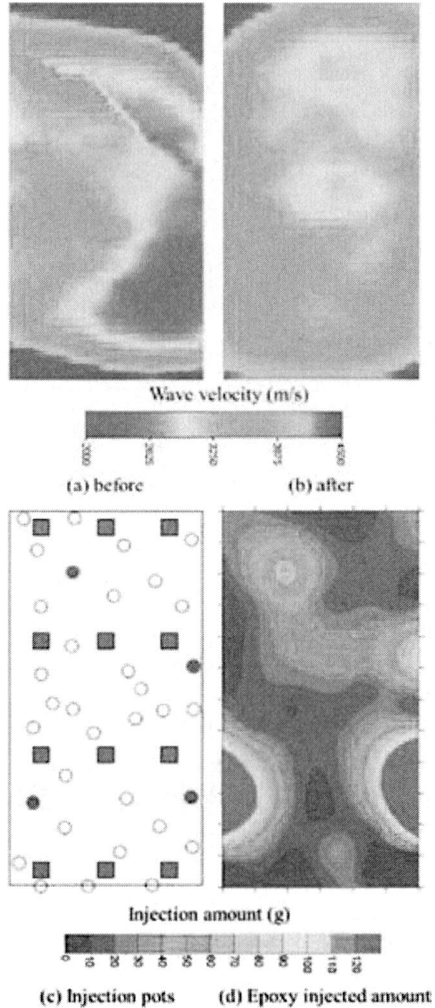

Figure 16.
Wave velocity and epoxy injection result (Side B).

Careful hammering at the steel bar head to prevent contacting the hole wall, the excited elastic waves were generated only from the bit hole bottom into the lining concrete, so that the excited signals were detected finally at sensors located on the concrete surface.

4.2.2 Propagation of waves in steel bar

The travel time along the steel bar was measured by two sensors as shown in **Figure 18**. AE sensor A records the excitation time at the head by using a steel ball of 15 mm diameter and the elastic wave travel time in the bar is calculated by

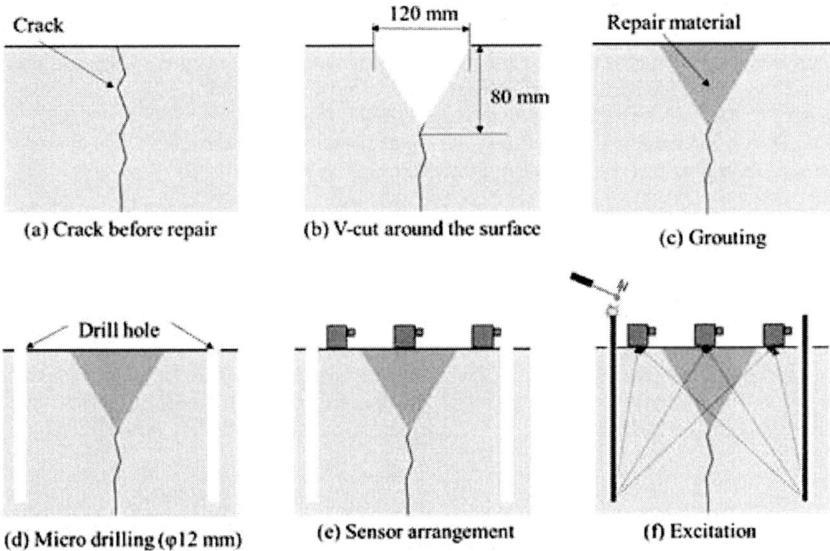

Figure 17.
Procedures of drilling and excitation.

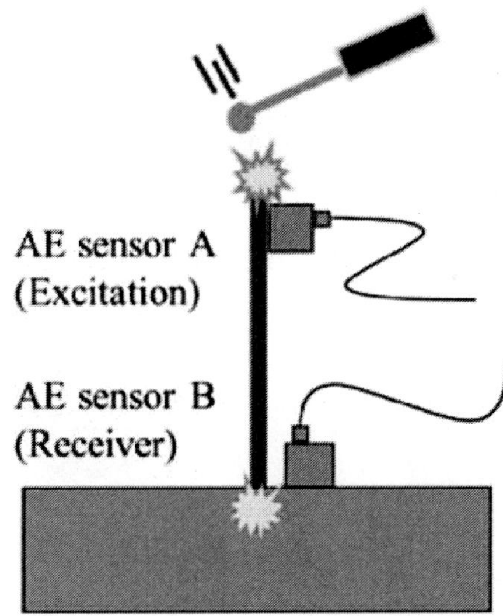

Figure 18.
Measurement method of travel time in the steel bar.

detecting the arrival time of the wave at AE sensor B. The arrival time difference is 69 μs as shown in **Figure 19**.

The dominant frequency of elastic wave excited by a 15 mm diameter steel ball is known as 19.4 kHz according to [12]. Considering a steel bar is used as wave guide, a frequency analysis was conducted for the waveforms observed at A and B.

Figure 20 shows the frequency spectra. Each guided wave is detected via the 38 mm length steel bar. The dominant frequency was observed at A for 22.5 kHz and the dominant frequency was 16.6 kHz at B. Since these detected frequencies are higher than the resonant frequencies of the steel bar, first flexural mode (1.1 kHz), second flexural mode (3.2 kHz), and third flexural mode (5.4 kHz), respectively, as a cantilever, the principal components of the waveform were assumed to be generated as compression wave excited by the tapping at steel bar head.

4.2.3 3D elastic wave tomography

The computation for wave velocity distribution in the targeted concrete wall was implemented by the tomography technique mentioned previously. **Figure 21** shows the 3D distribution of wave velocities and **Figure 22** shows them at cross section A.

Although the triangle-shaped (dashed line) repair area has high velocity on the surface, the V-shaped low-velocity area is observed toward the bottom, whereas high-velocity zones exist at the left side of the specimen. The high velocity may

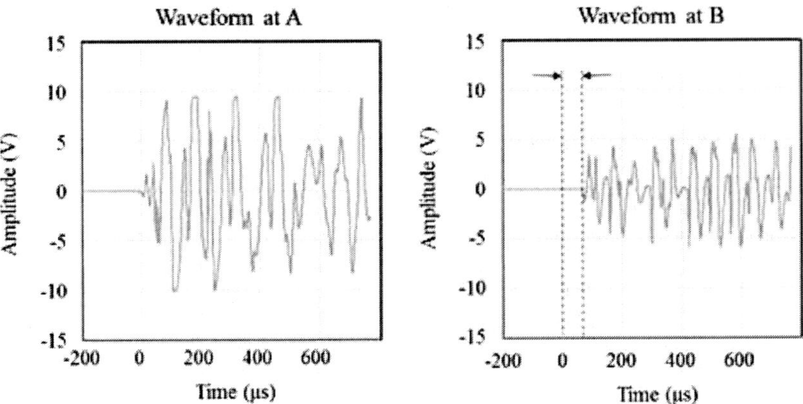

Figure 19.
Waveforms at excitation and receiver.

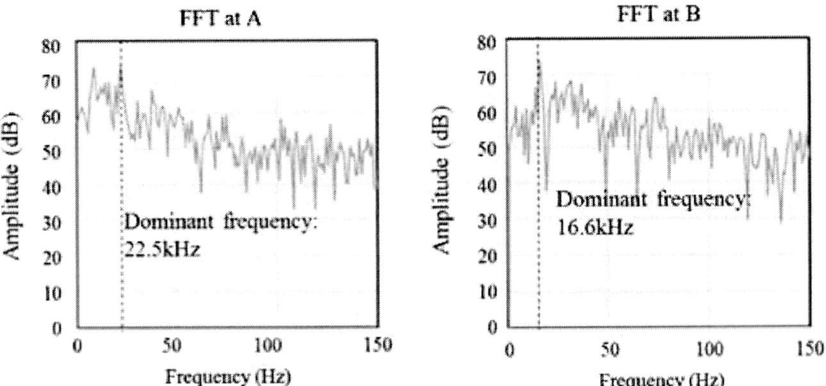

Figure 20.
Frequency spectra of waveforms at A and B.

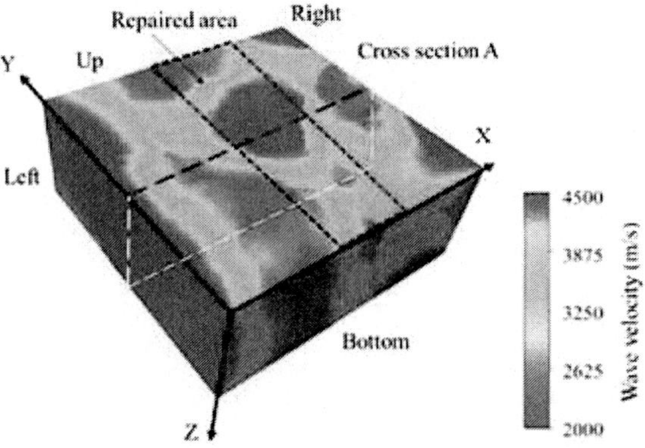

Figure 21.
Distribution of wave velocities in 3D.

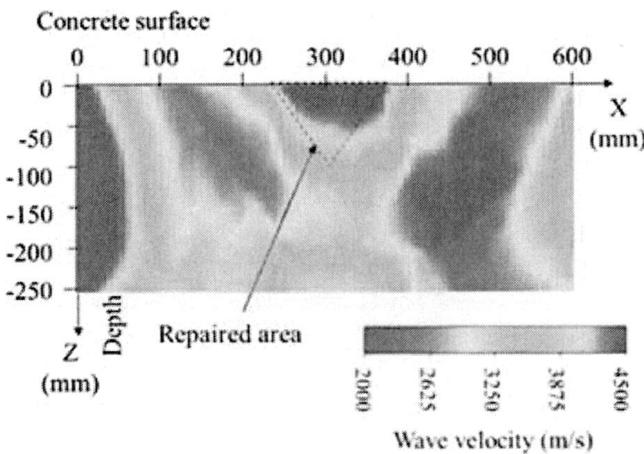

Figure 22.
Distribution of wave velocities at cross section A.

indicate the intact condition of the original concrete quality, because the interested area is enough far from the repaired area.

The repaired part is denoted by high velocity, in **Figure 22**, meanwhile the original concrete surrounding the patched area remarkably shows low velocity. The V-shaped area with low velocity underneath the repaired part could be potentially damaged by the chipping work for concrete removal. This is generally known and described in concrete surface treatment guideline prior to repairs and overlays [13, 14]. Further investigation is needed for the consideration in the influence of the hammer drill impact on damage to the concrete behind the removal zone.

4.3 Slab

4.3.1 Velocity distribution of AE tomography

Figure 23 shows the results of AE tomography, before and after repair by means of the crack injection. Results show that in all the slab panels, the velocity after

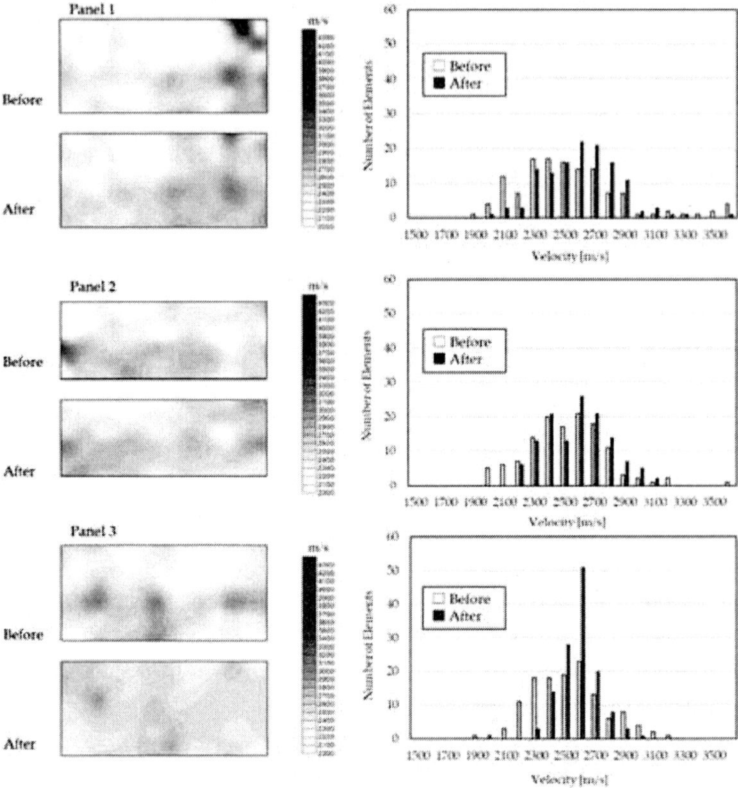

Figure 23.
Results of velocity distribution before and after repair.

repair exhibits increase compared to that before repair. Further quantitatively, the histograms of velocities obtained in all the elements are shown on the right side of the figure. For all slab panels, it is evident that the velocities at the elements clearly shift to the higher regions after repair. Due to the effectiveness of injected material in filling cracks and defects, detours and dispersions in the propagation paths of elastic waves are so eliminated that apparent velocities are increased.

All results imply that the velocity distribution obtained by the AE tomography method has a good potential to be an indicator for ascertaining the filled situation of injected material in a concrete slab. It is confirmed that the velocity for concrete which is not damaged shows about 3500 m/s to 4000 m/s. In some areas, however, velocities of about 2600 m/s are observed even after repair. This is because injected material might not be injected well into continuous cracks, independent air bubbles could be present due to the use of the air-entraining agent, and fine cracks at the interface between coarse aggregate are nucleated due to the alkali-silica reaction. As a result, there exists a possibility that the velocity recovery does not reach to the satisfactory level even after injection. On this issue, we plan to carry out a material test in the laboratory for confirmation.

4.3.2 Relationship between velocity and injection amounts

Design of the injection amount for the crack injection method could be based on the estimation of the crack widths, the depths, and the length measured. It is recognized that there exists no reasonable relationship between the amounts of designed injection and actual injection. Thus, an attempt to examine the amount of injected material is made from the results of AE tomography before repair.

It is considered that the amount of injection should increase, depending on the extent of damage. Namely, if the degree of damage is small, the amount should decrease. In addition, if the damage is less than a certain degree, the injected material may not work well on the damage. On the other hand, if the elastic wave velocity could reflect the degree of damage, a correlation should be evident between

V_P (m/s)	Quality
>4570	Excellent
3660–4570	Fine
3050–3660	Acceptable
2130–3050	Unacceptable
<2130	Poor

Table 2.
Quality indicator (Whitehurst).

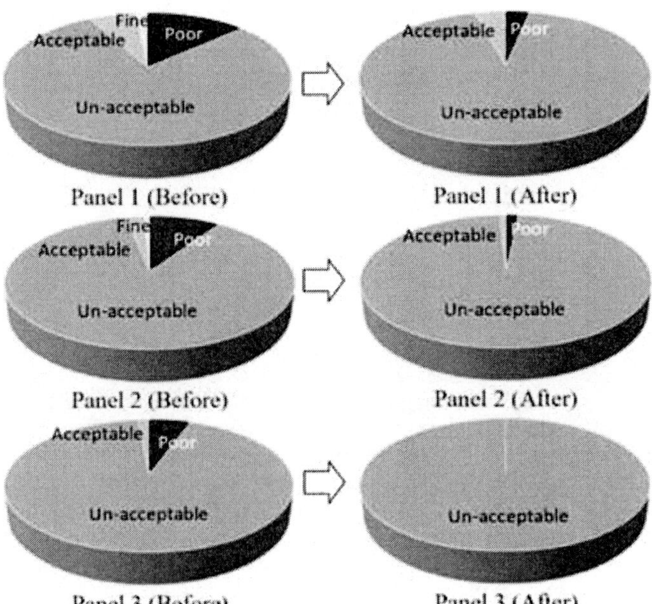

Figure 24.
Area ratio by quality before and after repair.

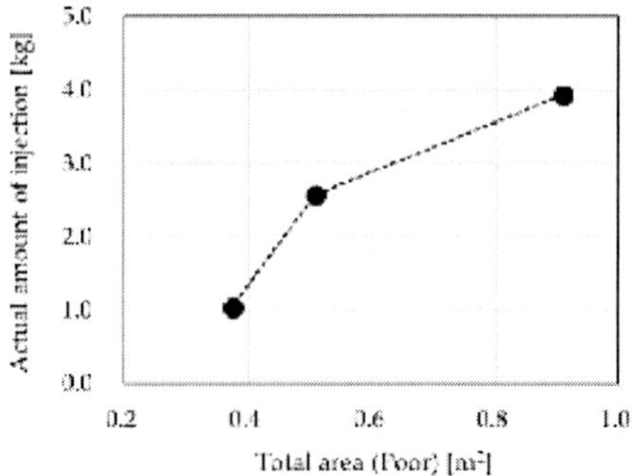

Figure 25.
Total area (Poor) vs. injection amount.

the amount of injection and the values of velocities. Thus, the velocities are classified into grades, as given in **Table 2**. These quality indicators are proposed by Whitehurst [15]. They were determined from the relationship between mechanical properties and P-wave velocity in concrete. Following these indicators, the qualities before and after repair of the panels are classified as shown in **Figure 24**. It is found that the number of elements with Poor decreases, while that of Unacceptable keeps almost the same from before to after repair. As discussed before, due to the presence of air bubbles and the damaged interface with aggregates by alkali-aggregate reaction, the recovery of the velocities may not be apparent. These results imply that the region where the injected material could improve the quality of concrete is mostly that of Poor. It suggests that the repair by means of injection is effective for comparatively major damage. **Figure 25** shows the relationship between total area of Poor estimated by AE tomography before repair and the actual amount of injection. As the Poor area increases, the increase in the actual amount of injection is clearly observed. Thus, it is possible to estimate the amount of injection before repair by carrying out the analysis using AE tomography.

5. Conclusion

Concrete pier, concrete wall, and slab were tested on the investigation on the internal damage assessment for the repair condition by applying elastic wave tomography and AE tomography. Determining the 3D velocity distribution, the repair effects of the epoxy injection method and the patch repair method were quantitatively evaluated. From the results, the following conclusions can be drawn in this study:

1. 3D elastic wave tomography technique can evaluate the penetration of repair epoxy injection material and qualify the repair effect with the amount of injected rexin. 3D tomography technique installed with single-side access drill hammering successfully visualizes the internal quality of concrete after the patch repair method based on the elastic wave velocity distribution.

2. The velocity distribution obtained by AE tomography can serve as an indicator for ascertaining the state of crack and void filling with injected material. A good correlation is found between the low velocity region before repair and the

amount of injected material. The results clearly show the potential for the AE tomography technique to be used as a method for estimating the performance of the crack injection method.

As mentioned previously, the RILEM committee was launched because innovative nondestructive inspection testing to qualify repair works is strongly required worldwide. We plan to continue studies based on the evaluation method using elastic wave tomography and accelerate its standardization.

References

[1] Damage assessment in Consideration of Repair/Retrofit-Recovery in Concrete and Masonry Structures by Means of Innovative NDT, Technical Committee IAM, RILEM; http://www.rilem.org/gene/main.php?base=8750&gp_id=347

[2] Kobayashi Y, Shiotani T, Shiojiri H. Damage identification using seismic travel time tomography on the basis of evolutional wave velocity distribution model. In: Structural Faults and Repair 2006 (CD-ROM); 2006

[3] Kobayashi Y, Shiotani T, Aggelis DG, Shiojiri H. Three-dimensional seismic tomography for existing concrete structures. In: Proceedings of Second International Operational Analysis Conference; 2007. Vol. 2. pp. 595-600

[4] Kobayashi Y, Shiotani T. Seismic tomography with estimation of source location for concrete structure. In: Structural Faults and Repair 2012, CD-ROM; 2012

[5] Asuae H, Shiotani T, Nishida T, Watabe K, Miyata H. Applicability of AE tomography for accurate damage evaluation in actual RC bridge deck. In: Structural Faults & Repair Conference, No.1743; 2016

[6] Kobayashi Y, Shiotani T. Computerized AE tomography, innovative AE and NDT techniques for on-site measurement of concrete and masonry structures. In: State-of-the-Art Report of the RILEM Technical Committee 239-MCM; Springer: 2016. pp. 47-68

[7] Akaike H. Markovian representation of stochastic processes and its application to the analysis of autoregressive moving average processes. Annals of the Institute of Statistical Mathematics. 1974;**26**(1): 363-387

[8] Zhang H, Thurber C, Rowe C. Automatic P-wave arrival detection and picking with multiscale wavelet analysis for single-component recordings. Bulletin of the Seismological Society of America. 2003;**93**(5):1904-1912

[9] Osawa S, Shiotani T, Kitora H, Momiyama Y. Damage visualization of imperfectly-grouted sheath in PC structures. In: 31st Conference of the European Working Group on Acoustic Emission, German Society for Non-Destructive; 2014. http://www.ewgae2014.com/portals/131/bb/fr1b3.pdf

[10] Sassa K, Ashida Y, Kozawa T, Yamada M. Improvement in the accuracy of seismic tomography by use of an effective ray-tracing algorithm. In: MMIJ/IMM Joint Symposium Volume Papers; 1989. pp. 129-136

[11] Kikusui Chemical Industries Co. Ltd. Inside Pressure Hardening. http://www.kikusui-chem.co.jp/pdf/products/catalog/cat_iph_3.pdf

[12] Sansalone MJ, Streett WB. Impact-Echo, Nondestructive Evaluation of Concrete and Masonry. Ithaca, NY: Bullbrier Press; 1997. pp. 29-46

[13] U.S. Department of the Interior Bureau of Reclamation Technical Service Center. Best Practices for Preparing Concrete Surfaces Prior to Repairs and Overlays; 2012

[14] American Concrete Pavement Association, Guideline for Partial-Depth Spall Repair, Concrete Paving Technlogy; 1998

[15] Whitehurst EA. Evaluation of concrete properties from sonic tests. In: ACI Monograph No.2; ACI; 1966

Index